Light Curve Modeling of
Eclipsing Binary Stars

E.F. Milone
Editor

Light Curve Modeling of Eclipsing Binary Stars

With 43 Illustrations

Springer-Verlag
New York Berlin Heidelberg London Paris
Tokyo Hong Kong Barcelona Budapest

E.F. Milone
Department of Physics and Astronomy
University of Calgary
2500 University Drive
Calgary, Alberta T2N 1N4
Canada

Cover figure adapted from Figure 3 of the chapter "Binary Maker 2.0 – An Interactive Graphical Tool for Preliminary Light Curve Analysis" by D.H. Bradstreet.

Library of Congress Cataloging-in-Publication Data
Milone, E.F., 1939–
Light curve modeling of eclipsing binary stars / [edited by] E.F. Milone.
p. cm.
Includes bibliographical references.
ISBN 0-387-97946-8. – ISBN 3-540-97946-8
1. Eclipsing binaries – Light curves. I. Title.
QB821.M55 1993
523.8′444 – dc20 92-21532

Printed on acid-free paper.

Production managed by Hal Henglein; manufacturing supervised by Vincent R. Scelta.
Camera-ready copy prepared by the contributors.
Printed and bound by Edwards Brothers, Inc., Ann Arbor, MI.
Printed in the United States of America.

9 8 7 6 5 4 3 2 1

ISBN 0-387-97946-8 Springer-Verlag New York Berlin Heidelberg
ISBN 3-540-97946-8 Springer-Verlag Berlin Heidelberg New York

Contents

Improvements in the Modeling of Eclipsing Binary Light Curves- An Introduction

This book represents the augmented proceedings of a two-session meeting on Improvements in Light Curve Modeling, a specialized topic in the field of close-binary stars, sponsored by Commission 42 (Close Binary Stars) and planned for two scientific sessions of that Commission during the International Astronomical Union's XXI General Assembly in Argentina at the end of July, 1991.

The nature of the modeling of light curves of eclipsing binary stars has not been discussed in a formal way for nearly two decades, since the advent of eclipsing binary modeling code. Then, the principal concern was how results of such operations compared to those from the trusted and widely-used Russell-Merrill technique. While still in use today, this venerable method is being replaced increasingly by more robust techniques which potentially can model all systems better, but especially close binary systems where proximity effects cannot be ignored. Better physics, it is fair to say, has won the day, and modern modeling techniques are clearly here to stay, but concerns can be raised about modeling philosophy, performance, and what improvements are needed and can be implemented in the near future. For example (one may ask), are there compelling physical reasons *not* to use a precision code to model the effective temperatures of the component stars, their compositional differences, or their transient characteristics? In other words, what should be the goal of a modeling program? More practically, there is the question of responsibility for and maintenance of the codes: in the future, who will upgrade them, port them to new platforms, issue 'bug warnings', and incorporate improvements? Should codes be made more 'open' so that changes can be introduced by different parties? If so, how? These questions are addressed in some of the papers, but most practitioners have already answered them for themselves, and demonstrate their improvements to, or applications of, the modeling methods and paradigms they have been using. The improvements, both realized and promised, are impressive.

Potentially superior determinations of fundamental and even transient stellar parameters can now be anticipated from the burgeoning developments in light curve analysis techniques and software and the availability of powerful, and sometimes dedicated, computing facilities. The new light curve analysis techniques include simulations of stellar bright and dark spots, streams, partial and complete stellar disks, prominences, and other features characterizing chromospherically active stars; modeling of polarization parameters. Spectroscopic modeling techniques associated with light curve analysis code include not only the use of radial velocities but also application of line profile simulations to model both velocity field variation across stellar disks, and the weighted effects of brightness asymmetries. Finally, translucent eclipsing agents such as stellar winds are being modeled with increasingly sophisticated methods. Progress in many of these areas can be recognized in the work described here.

In addition to the actual content of the analysis and synthesis programs, software technique philosophies now span the range from interactive to pure batch operations. These philosophies reflect different ways of investigating parameter space, which must be thoroughly explored nevertheless to insure the uniqueness of the solution, if indeed it is uniquely determinable from the available data set. The limits for solution of a given body of information is a lively current topic of research in its own right.

Among batch operations are multiple runs of parameter fittings, a mode specifically discouraged by Wilson and Devinney. The software technique which is embraced may in turn affect, or be affected by, the type of hardware on which the analyses can be carried out. Supercomputers provide superb platforms with which to examine and to determine the absolute minima of parameter space, but dedicated workstations may be fast enough and provide enough temporary storage to adequately sample parameter space at much lower cost. A solution to the fundamental problem of modeling - determination of uniqueness of the 'best' parameters from optimization in a local region of parameter space - is certainly closer because of these powerful tools. The whole question of the advantages and disadvantages of the different modeling techniques, philosophies, and tools had not been aired for considerable time, and, for this reason alone, deserved to be discussed in as general a forum as the community can provide.

The meeting was originally scheduled to be held during the entire morning (0900:12:30) on July 31 in the San Martin Cultural Complex. However, that morning the Union was confronted by a scene of billowing smoke, firemen and equipment, and a flurry of activity at the meeting site. Because of smoke damage, the venue was no longer usable; our sessions were postponed for seven hours and truncated to one session. By agreement of our *SOC*, only those speakers who had not planned to attend Colloquium No. 151 in Cordoba were asked to present their papers at the General Assembly (with one exception). It is a tribute to the organizational skill of the IAU Local and Scientific Organizing Committees that we were able to begin our proceedings by 1600 in a comfortable cafe, and a tribute to the generosity of the LOC and SOC of Colloquium No. 151 that we could conclude our meeting in the attractive Salon Turquesa of the Gran Hotel Dora in Cordoba on August 7 and that most of the poster papers, which were to be put on display on July 31, were indeed presented in this form in the posters area of the meeting of Colloquium 151.

I had intended to open the proceedings by making a few remarks in Spanish, but due to the time constraints could not include them in my remarks at the time. I can do so now, and add the full version of my briefest remarks introducing the topic and the first speaker:

'¡Buenos días y bienvenidos!

Es un placer participar de este Congreso de la Uníon Astronómica Internacional aqui in Buenos Aires, y en nombre del Comité de Organización Científico de la Sesión de Modelado, deseo expresarles nuestro agradecimiento. Las primeras dos sesiones de esta mañana estan dedicadas a una reunión cientifica auspiciada por la Comisión 42 (Estrellas Binarias Eclipsantes). La reunión se titula "Mejoras en el Modelado de las Curvas de Luz de Estrellas Binarias Eclipsantes" y es la primera reunión en muchos años. Como peuden ver, nosotros tenemos un interesante programa (los horarios estan su disposición aquí). Ahora desearía presentar (en inglés) al primer orador. Les ruego sépan disculpar mi español.'

The actual introduction involved a brief description of the shortened program, the rearranged schedule, a strict warning to the speakers that the timetable would be adhered to regardless of the state of completion of the presentation, and the promise that somehow or other, the remaining papers would be presented in Cordoba. The actual program, presented at each of the two venues was as follows:

Café Konzert, Buenos Aires

16:00 E. F. Milone: *Introduction I.*

16:05 R. E. Wilson: *From Here to Observables: Beyond Light and Velocity Curves*

16:20 D. Terrell: *Mass Transfer in Algols: Computation of Observables*

16:30 J. Kallrath: *Gradient Free Determination of Eclipsing Binary Light Curve Parameters - Derivation of Spot Parameters Using the Simplex Algorithm*

16:40 R. Kurucz: *Improvements in Stellar Atmosphere Models*

16:50 W. Van Hamme: *The New Wilson Reflection Treatment and the Nature of BF Aurigae*

17:00 K.-C. Leung and D. Q. Zhao: *Treatment and Interpretation of Symmetry in Light Curves* (presented by Leung).
17:15 General Discussion I.

Hotel Gran Dora, Cordoba

19:00 E. F. Milone: *Introduction II.*
19:10 E. Budding: *Review of Poster Papers*
19:30 General Discussion II. Presentation by A. M. Cherepaschuk.
19:40 E. Guinan: *Disk System Modeling: Models for ε Aurigae*
19:50 P. Etzel: *Physical Applicability of the Nelson-Davis-Etzel Model and Status of the EBOP Computer Code*
20:00 C. R. Stagg & E. F. Milone: *Improvements to the Wilson-Devinney Code on Computer Platforms at the University of Calgary* (presented by Milone).
20:10 General Discussion III.

Both meetings were time-constrained: the first by another meeting arranged for the Café Konzert; the second by a consideration that the rental of the room was costing someone (and it was not entirely clear at the time who that was) money. The strange bifurcation of our sessions had the unusual benefit of enlarging the audience who were able to participate in the proceedings; it had the disadvantage of not giving each audience the benefit of having heard all of the speakers. Hopefully, this volume will fill in some of the blanks for those attending while making a wider audience aware of the state of light curve modeling.

Several of our initial invited speakers had to withdraw for personal or financial reasons or because of conflicting schedules. Dr. Linnell could not attend because of serious family illness; Dr. Hill had an observing conflict. Walter Van Hamme, unable to attend the

Cordoba meeting, was prepared and able to present his poster paper orally in Buenos Aires. Several speakers were able to provide manuscripts although they themselves were not able to attend. In every case, the authors were instructed to present the written presentations they wished they had said, rather than what they did say. All papers were subject to a refereeing process for improvements. It should be noted that the comments and questions which follow each paper were made at the time of presentation and may deal with matters now treated within the present text. The shorter poster papers were reviewed orally by E. Budding, and a brief version of his remarks are presented here. The division of the meeting into two sections creates a difficulty in ordering the written papers: should one organize them as presented or by subject matter? I have chosen a middle road: the Buenos Aires papers are presented more or less in the order in which they were given. These are followed by oral and poster papers which were given in Cordoba except for the Stagg-Milone paper, which was the very last paper presented in Cordoba, but is more closely related to the other papers in the Buenos Aires part of the program. The papers of invited speakers who could not attend the meetings have been inserted at appropriate places in the program; poster papers which were described or presented in Cordoba, and recorded remarks made during the second session, including those of Dr. Cherepashchuk, appear in the latter part of the proceedings. The order of the program as it actually unfolded has been indicated above. The placement of Dr. Budding's poster review prior to several speakers was dictated by the indisposition of Dr. Etzel, who was suffering from a bad cold, and had to be awakened from a sickbed! The Samek and Hube poster paper cited by Budding was not originally part of the meeting, and its authors have decided against submitting a written version of their paper. Further comments about the organization and accomplishments of the meeting can be found in the concluding remarks.

- E.F. Milone

The University of Calgary, July 1992

From Here to Observables: Beyond Light and Velocity Curves

R.E. Wilson
Astronomy Department, University of Florida, Gainesville, FL 32611, USA

I. Applications of "Light Curve" Models

About 20 years ago, several computer models for binary star light curves appeared. At the same time it became possible to fit the models to light curves impersonally by the least squares criterion. With development of those and other models came the end of treating light curves of close binaries by graphical trial and error methods, with overly simple or non-existent physics. Improvements over the years have made it possible to fit radial velocity curves, to make **simultaneous** solutions of several data sets, and to handle many physical effects which are far beyond the old capabilities. It would require excessive space to describe such a program, but a basic overview of one of the programs can be found in Wilson (1979, 1990, and 1991). The computer models not only are physically realistic, but are sufficiently flexible to allow ready addition of new features. A good model includes the various inter-connected phenomena which once made binary systems intractable, for example tidal and rotational deformation, mutual heating, eccentric orbits, gravity darkening, limb darkening, and spots. Persons who are familiar with these computer models can add features where needed for unusual binaries. It would be a missed opportunity to have all of this capability and use it only for light curves, when it could be used to predict virtually any integral property over the surfaces of the stars, such as spectral line profiles, narrow band indices, and polarization measures. Existing "light curve" models are well suited to the realization of these objectives because they already contain the required physical model, the properties of which can be varied to match properties of real binaries. Many numbers needed to compute, for example, spectral line profiles or polarization curves, already exist within the computer and only relatively simple further computations are required to synthesize the new quantities.

However considerable development is in order if many kinds of observables are to be generated efficiently with one program. This paper is intended as an outline of ideas which can lead to generalized and efficient binary star "observables" programs. It particularly will address the problem of the interface between theory and observation. The availability of computational models and analytical procedures should stimulate observers to augment the present small accumulations of polarization curves, narrow band index curves, and absorption line profiles.

II. Observables Other than Light and Radial Velocity Curves

Consider four traditional kinds of observations:

1. polarization curves
2. narrow band index curves
3. photospheric spectral line profiles
4. emission lines from circumstellar gas

These are greatly under-utilized at present (analyzed only sketchily and qualitatively), and for that reason the first three kinds are made only infrequently. Naturally an observer will focus on types of observations which can yield hard numbers. Since few of these observations have been published, persons who develop models and analysis have not paid much attention to them — why develop capabilities to model and analyze non-existent observations? The four areas are being explored by R. E. W. and collaborators and also by some other groups. At this writing, we are doing least squares fits of combined limb and circumstellar polarization curves, which are discussed in Section III. Ideas on modeling narrow band index curves are in Section IV.

Our work (J. D. Mukherjee, G. J. Peters, and R. E. W.) on photospheric line profiles is primarily for the purpose of measuring rotation, and we expect that even latitude dependent differential rotation will be accurately measurable in favorable cases. The primary need for good rotation measures is to test structural-evolutionary models of stellar rotation. The rotational state of a star is, to a limited extent, predicted by theories of tidal evolution, mass accretion, mass loss, and angular momentum re-distribution (e.g. Zahn, 1977; Scharlemann, 1981, Tassoul, 1987; Tassoul and Tassoul, 1990). Much strengthened tests of current theories are required in order to place the various proposed mechanisms into perspective. Proper understanding of rotation in single and binary stars is critical to successful computation of stellar evolutionary sequences. Our fitting of profiles (observed by G. J. Peters) should provide much improved rotational statistics for accreting stars in close binaries, and eventually in similar statistics for single stars. In the subjective graphical results available until now, it is impossible to estimate, even roughly, how much of the variation from one publication to another is due to subjective technique, how much to noise, and how much to real astrophysical changes. Standard error estimates are missing, and the models are simple limb darkened spheres. Our computed profiles are integrated over the surfaces of Wilson-Devinney (hereafter WD) model stars and include damping, thermal and turbulent Doppler motions, instrumental broadening, and phase smearing, as well as all of the binary star effects of the WD model. Progress to date on photospheric line profiles, including rotation measures for 6 Algol type primary stars, is summarized by Mukherjee, Peters, and Wilson (1991). Other recent work on profiles is that by Gies and Wiggs (1991), Eaton, Hall, and Honeycutt (1991), and Richards (1992).

Our programs to generate and fit emission line profiles are at an intermediate stage of development. NASA has collected enormous reservoirs of close binary star spectral energy distributions (SED's) observed from space. However analysis of this potentially valuable resource has been primarily qualitative, consisting largely of the noting of features, but not quantitative modeling and analysis. We have a computer model of the flow and radiation of gas in interacting binaries (Terrell and Wilson, 1992). Here the WD model provides only an added source of continuum radiation and the illumination of the gas, which are minor roles. However, extension of the idea of a complete model into the realm of circumstellar gas continues the new tradition of treating the general case with a general model.

Flow computations have a long history (Prendergast and Taam, 1974; Lubow and Shu, 1975; Lin and Pringle, 1976; Whitehurst, 1988a; 1988b) and there have been contributions on the radiative properties of optically thin gas under conditions typical of close binaries (e.g. Krolik and McKee, 1978, Drake and Ulrich, 1980), but the two parts of the problem have not previously been put together. Our interacting gas flow model includes viscosity and pressure and, while it utilizes mainly previous ideas of orbital mechanics and physical interactions, it has many new efficiencies which make computation much faster than was possible in earlier work. The intrinsic speed makes it practical to use large numbers of gas elements and to vary parameters on a scale not previously attempted. However the newness here is in doing the complete problem all the way to synthesized spectrometry, with Doppler broadened and shifted radiation. At present we use a radiative computer program by S. A. Drake, although later we expect to develop a new one. Because our gas dynamics program keeps account of the absolute quantity of transferred matter, it should be possible to infer well-founded mass transfer rates (dm/dt). Evolutionary models predict dm/dt at various evolutionary stages, but observational estimates have been little more than guesses. We shall be able to model timewise progression of disk behavior for comparison with space observations over many orbital cycles and produce meaningful tests of understanding.

A common problem for all four observables is a lack of tradition of publishing observations. It is not clear why so many papers appear in which only graphs or verbal descriptions of observations are given. This is especially disconcerting in cases where only small or modest-sized tables of numbers would suffice to record the observations in useful form. The result is that large expenditures of time, effort, and money go mostly to waste because, despite assurances, most observations have been mislaid, thrown away, or left in some unusable form within a year or two. Observers move their offices, become enmeshed in other work, and have finite lifetimes. Data banks are somewhat better than no recording at all, but have enough shortcomings to take up a paragraph of their own, and are not a satisfactory solution. An efficiently run and otherwise idealized data bank might seem to cover the need, but would be needed only for large data sets - most of the required tables are too small to bother sending to a data bank. They just need to be published. This impediment to progress occurs in only a minority of light curve and radial velocity papers, but is the rule rather than the exception for many other kinds of observations.

Observations from space will be needed. While SED's can be and have been made within existing space programs, the four kinds of required data are mainly for future instruments and programs. Space observations are needed for different reasons among the four observational areas. For circumstellar polarization and for SED's arising in circumstellar gas flows, the primary reason is the episodic nature of the flows, which need to be followed over at least several orbital cycles, uninterrupted by the diurnal cycle or by star risings and settings. An almost equally important reason with regard to SED's is that most of the interesting spectral lines are in the far ultraviolet, and are not observable from the ground. Space observation of limb polarization is particularly critical because the effect is small in the visible spectral region but increases very strongly toward short wavelengths (Collins, 1970). For narrow band index curves, the main impetus for space observations is in the need for maximum achievable accuracy. While selected binaries should have index curves with reasonably large amplitudes, many others will have only modest or small amplitudes, yet could yield interesting results from high precision space observations. Accuracy, will be a prime concern even for the relatively large amplitude stars. For rotation from photospheric line profiles, the need for space observations becomes apparent when one actually goes through the analysis. It results from the difficulty in finding unblended lines. Significantly blended lines are not useful for this purpose, and the blending can be with lines of the same star or of the other star. While blending problems are usually even worse in the UV than in the visible, space observations increase opportunities for locating unblended lines, mainly because of access to the infrared. The best strategy is to measure rotation from as many apparently unblended lines as possible, and demand consistency.

III. Circumstellar and Photospheric Polarization

The polarization effects which have received the most attention in binary star work are polarization due to scattering from circumstellar gas and photospheric limb polarization. Limb polarization aroused much interest after Chandrasekhar (1946) treated the pure electron scattering case, which is approximated in extremely hot stars, and pointed out that eclipses could break disk symmetry and reveal the effect to observers. Refinements by Code (1950), Collins (e.g. 1970, 1972), and others predicted only very small eclipse polarization effects, but left the door open for larger effects possibly arising due to departures from canonical stellar atmosphere models (Harrington, 1969). After unsuccessful attempts by numerous observers to detect the limb eclipse effect, Kemp, et al. (1983) announced success for Algol. This result has not been uniformly accepted as genuine limb polarization for several reasons (private communications), but our modeling indicates that limb polarization causes a significant part of the rapid variation within primary eclipse. Our methods and Algol results will be published soon (Wilson and Liou, 1992, hereafter WL). A sketch of the model and solution scheme follows below.

In order to maintain flexibility and thus facilitate generality, we follow a somewhat different path from other polarization models. All of the included astrophysical effects are **incorporated within the model**, rather than removed from the observations. Often one reads discussions of how to correct observations for some effect which is not in a model, but such corrections should be avoided whenever feasible. The corrections are always linearized and do not combine correctly with other effects outside the linear regime. In approximate theories, effects will not combine correctly even within the linear regime. We generate the Stokes quantities Q and U, first in the absolute versions for the individual sources, and then combined into their relative versions, which an observer will record (we do not use the term "Stokes parameters", since in essence they are directly measurable quantities, not parameters). It is important to compute Stokes Q's and U's in full analogy with those measured by an observer, so that direct comparison becomes a straightforward process. Thus we combine absolute numbers into relative Q's and U's, which means that we form weighted means, with the weights being observable fluxes from the several unresolved sources. A more specific discussion is contained in WL.

For polarization by circumstellar matter, relations between source geometry and Stokes quantities can be found in a widely cited paper by Brown, McLean, and Emslie (1978). In a binary star system there are two sources of illumination, which we currently treat as points (for illumination, but not for eclipses). We maintain flexibility by treating only individual clouds as scattering elements, rather than the rings and disks of most other work. Extended structures are made up of distributions of optically thin scattering clouds, each characterized by a volume, an electron density, and a location (x, y, z) in the rotating coordinate frame of the binary. Two questions arise in connection with this strategy. First, since polarization integrals over rings, disks, and other figures have been derived and published in several papers, why should we not use those integrals where they can be applied? Second, since virtually all circumstellar matter is in continuous motion, what is the value of putting clouds at fixed locations? Both of these questions are answered if one realizes that the present model is intended as a stepping stone to a more physical model. A hydrodynamic flow program, such as that described by Terrell and Wilson (1992), can serve for polarization as well as for spectral energy distributions. Most of the trouble of constructing a polarization program occurs in setting up and verifying coordinate systems which are fully consistent with the conventions of observers. Once this has been done, it is rather simple to couple the polarization program to a hydrodynamic flow program – the coordinates, etc. of the hydro elements are just assigned to the polarization program coordinates at each time step. Thus the natural geometrical elements are point-volumes, not rings or disks. However a simple fixed-location polarization program can serve a useful intermediary purpose by identifying the gross features of circumstellar gas distributions such as an accretion stream and an orbiting ring. Then by combining a polarization program with a gas flow program, one can make polarization predictions with far more physics than before. In concert with modeling of spectral energy distributions, this will test theoretical flow models. A major problem in circumstellar polarization work has been the large number of free parameters in arbitrary geometrical models. Introduction of physics greatly reduces the number of free parameters, and opens a new window on flows.

Fitting of limb polarization can give a new handle on binary star parameters because the detailed polarization behavior is sensitive to exact eclipse circumstances. The custom in other work has been to compute limb polarization signatures on the assumption that polarization occurs only on a thin ring extremely close to the limb. As shown by WL, this gives schematically correct but not accurately correct polarization curves. WL also show the effects of eclipses of the accretion stream and of rescaling in brightness due to the mutual eclipses of the stars, both of which are intermingled with the limb eclipse effect. For now WL use a scaled-down pure electron scattering theory (Chandrasekhar, 1946) for the local aspect dependence of polarization, which is adequate for exploratory purposes. Eventually we expect to develop a full stellar atmosphere treatment.

WL have compared combined limb and circumstellar polarization curves, by the least squares criterion and also subjectively, with the polarization curves of Algol observed by Kemp, et al. The findings are too extensive to summarize here, but one of the early subjective fits is shown in Figure 1. The least squares fitting is by the method of differential corrections, and is helped by the fact that partial derivatives of the Stokes quantities with respect to the polarization parameters can be

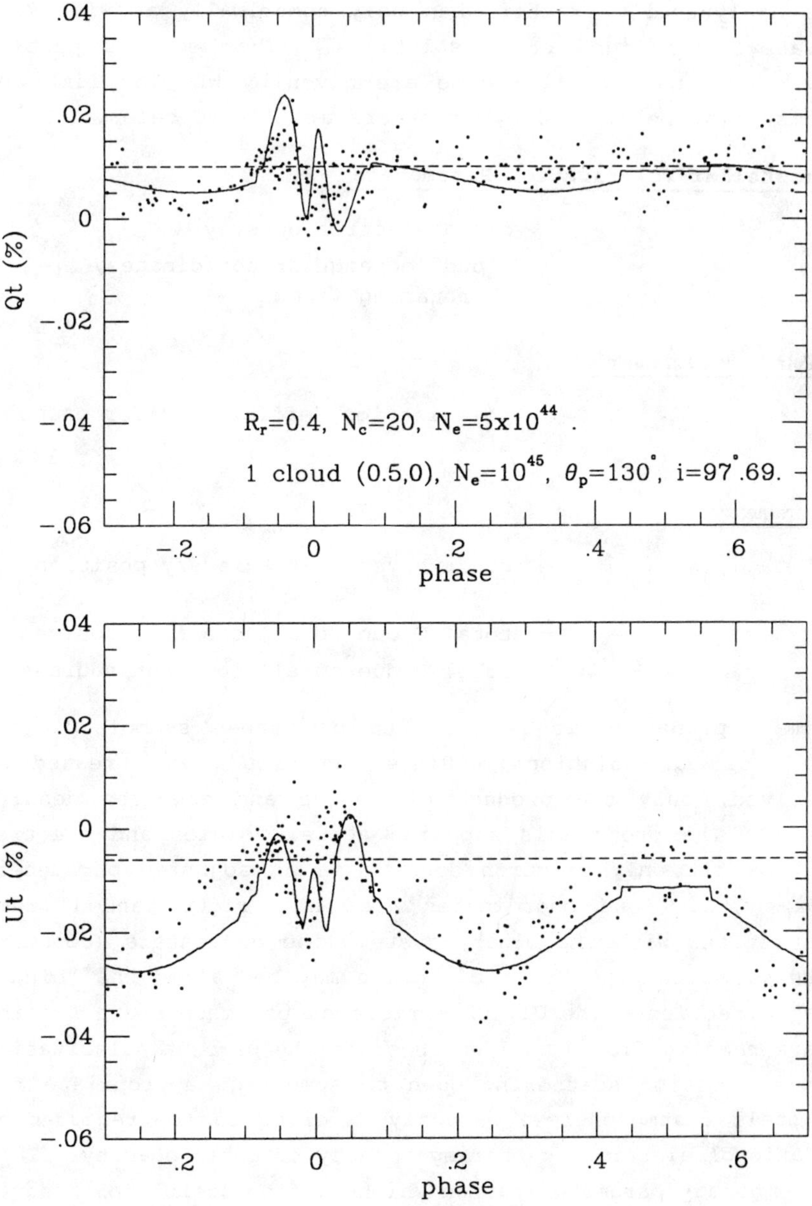

Figure 1. Preliminary subjective fit to Algol polarization curves, including both limb and circumstellar polarization.

found analytically, rather than only numerically as with most derivatives in light curve solutions. The required partial derivatives and overall scheme are given by WL. The limb and circumstellar polarization parameters are listed below.

Circumstellar Polarization Parameters

n_e	cloud electron density
x, y, z	cloud rectangular coordinates, co-rotating frame

Limb Polarization Parameters

C1, C2	polarization "efficiency" for each star

Joint Parameters

θ_{pole}	binary system +z axis sky position angle
Q_k	total Q due to all constant sources
U_k	total U due to all constant sources

Some explanation is required to put these symbols in the context of our solutions. Since the clouds are treated as unresolved, only the product of volume and electron density matters. The program is supplied actual volumes and electron densities but only electron density is an adjusted parameter. The x, y, z cloud coordinates are in a right handed frame which rotates with the binary system line of centers (counter-clockwise motion). The +z direction may be called the "binary north" direction. The C1, C2 efficiency parameters are scaling factors ranging from 0 to 1 which scale the percent polarization of pure electron scattering down to something appropriate for real stellar atmospheres. Roughly C1 or C2 can be regarded as the ratio of electron scattering opacity to total opacity. This is a temporary parameterization which we find useful for program development. It will be replaced by a more direct stellar atmospheres computation as soon as practicable. The position angle of binary system north is described counter-clockwise from the north direction on the sky. It is the only one of our polarization parameters which is common to the circumstellar and limb effects, and it therefore serves the very important purpose of tying the two effects together. That is, solutions must produce one binary-north position angle which satisfies both effects. The Q_k and U_k constant Stokes quantities represent the Q and U sums of all constant effects, such as interstellar polarization, constant binary system polarization, and residual instrumental polarization which has not been removed.

Useful information results from applying the polarization model to the Kemp, et al. (1983) Algol observations. We confirm the finding by Kemp, et al. that Algol orbits clockwise on the sky, and with approximately the Kemp, et al. position angle. Those observations are presently unique in showing both limb and circumstellar variation quite clearly, and it turns out to be impossible to represent both effects with counter-clockwise motion. Limb and circumstellar Q's and U's can be satisfied separately for counter-clockwise motion, but the position angles do not agree. The fast variation in primary eclipse is only partly due to the limb eclipse effect, with other important causes being the re-scaling of the observable (relative) Stokes quantities due to the changing brightness of the primary star, and eclipse of the (secondary to primary) accretion stream. With all of these effects mixed together, it is crucial to add the Q's and U's of separate sources properly. The specifics of that computation are given by WL. In differential corrections solutions, WL have estimated the location of the accretion stream as well as the number of free electrons in the stream (Figure 1 is from a subjective first cut and is superseded by the results of WL).

A limb polarization model can give a new handle on ordinary binary star parameters because the detailed variation responds to eclipse geometry. The differential corrections solutions of WL are a first step in dealing with eclipse polarization, in which the geometry is assumed to be known from a light curve solution and only polarization parameters are adjusted. The next obvious step is to fine tune the geometry (inclination, radii) by developing an impersonal fitting program which operates simultaneously with light curves, polarization curves, and perhaps with other types of data. In parallel with this effort, one could make a program to fit circumstellar polarization by interfacing with a gas flow program, as mentioned above. Finally, those two programs can be merged into one which analyzes light, polarization, and other curves as they are produced by both photospheric and circumstellar processes.

IV. Narrow Band Index Curves

Narrow band indices of the type considered here were devised to measure spectral line strengths **photometrically** and thus bypass the use of photon-inefficient, expensive, and cumbersome spectrographs (most light goes into other orders of the grating, especially the zero order). The idea is to measure relative flux through two narrow band filters, one narrower than the other, with both centered on the wavelength of a spectral line so that the missing absorption line flux is a larger fraction of the signal through the narrow filter than through the wider one. The ratio of narrow and wide signals (normally expressed logarithmically) then can be calibrated against line equivalent width. Once the calibration has been done, we have a fast way to measure line strength which is essentially unaffected by instrumental sensitivity drift and by atmospheric extinction. The freedom from sensitivity drift is very important for space observatories, which are difficult to recalibrate after being placed in orbit. A properly designed photometer can achieve much better accuracy (in magnitudes) for such an index than is possible for flux measurements. The most commonly used such index is the β index for the second Balmer line of hydrogen. Information on the β index can be found in Stromgren, 1966, Crawford, 1973, Crawford, 1975, Bell, 1971, and many other papers.

Narrow band indices have mainly been used to estimate luminosities of stars. Seldom have observers measured the **variation of narrow band indices with phase** for binary stars

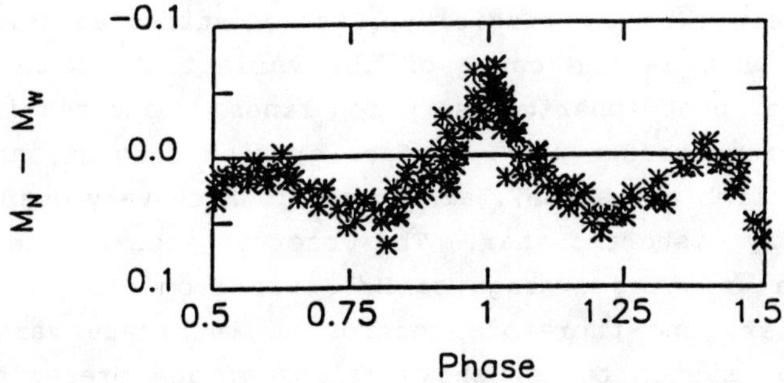

Figure 2. Hβ index curve of V1010 Ophiuchi observed by Corcoran, Siah, and Guinan (1991).

– after all, what could one possibly learn? Potentially, for binary stars with strong tidal distortions or radiative interactions, one can learn a great deal. Figure 2 shows the Hβ line curve of hydrogen for the near-contact binary V1010 Ophiuchi, as published by Corcoran, Siah, and Guinan, 1991. A glance at this curve shows that it contains considerable information – the amplitude is much larger than the scatter, there is substantial harmonic content (not just a simple sine wave), and it has to be cleanly periodic since it has been folded together from observations over many cycles. The standard thinking about Hβ variation in binaries is that it results from emission line fluctuations in circumstellar gas, but that cannot be the case for V1010 Oph because the stars are almost touching, so there is no room for circumstellar gas. Gas surrounding the entire system could produce variations, but not with the binary orbit period. In fact, there are many reasons to exclude circumstellar activity here (wrong periodicity, lack of strong emission lines, etc.), so what is the cause of the variation? Note that the strengths of photospheric absorption lines depend sensitively on effective temperature (T_e), surface gravity (log g), and aspect (center to limb variation), all three of which vary over the disk of a tidally distorted star. The observer sees a line strength which is a weighted average of Hβ lines from all parts of the visible disk, and therefore records an intricate variation in time as the system turns, shows eclipses, and presents various parts of its surface. Thus we see an interesting curve, as in Figure 2. However to learn anything about the binary from such a curve, we need a capability for synthesizing the curve's theoretical counterpart, and we need a procedure for estimating the parameters of a model by matching observed and computed curves. However it is not only for the Hβ line, but for many potentially useful lines, and it is far too early even to guess how many spectral lines will prove important. It is not only for binaries like V1010 Oph, with both stars having large tidal distortions, but for many binaries in which only one of the stars has a large tide, and for fully overcontact binaries with the stars sharing a common envelope. It is for binaries in which the distortion is due not to tides but to rapid rotation. It is also for binaries in which line strength varies because of local heating from the companion star. In short, the idea is potentially applicable to a good fraction of the thousands of systems in our catalogs of very close binaries.

What needs to be done is to incorporate local line profile information into a full binary star computer model (a "light curve" model) and integrate the product of local radiative intensity and bandpass response over the wavelength ranges of the narrow and wide filters of the β index, or other such index. The binary star program then must perform a flux integration over the visible disk of each star. The result will be the synthesized counterpart of an observable narrow band index. As the model binary orbits, this procedure will naturally produce the time-wise variation of the index, for direct comparison with observations. To do all this, line profiles must either be computed from a stellar atmospheres program or calibrated through observations of single stars. We intend to follow a combined approach to line profile calibration, utilizing both atmosphere programs and observations. With regard to observing index curves for binaries, it appears from several private communications that such observations are occasionally made, and that a few persons even do this routinely along with other photometry, but do not always publish the results. Of course, existing β index curves are not likely to exist preferentially for good analysis candidates, such as V1010 Oph.

The parameters of index curve modeling are in part the same as those of light curve modeling, but are more absolute in character because log g contains information on the absolute mass and absolute length scale (it measures M/R^2). The example of V1010 Oph suggests that we may have a new way to discriminate fine effects. Whether V1010 Oph is overcontact, semi-detached, or detached has been controversial. Note that line strength, integrated over visible disks and varying with phase, should be sensitive to surface morphology, such as the presence or absence of connecting necks. Note also that index curve analysis can be done by the same impersonal least squares algorithms which have proven so powerful for light curves. Some ideas about solving index curves follow in the next section.

V. Parameter Adjustment

Development of binary star computer models to include prediction of the quantities discussed above, and perhaps other quantities, should stimulate increased activity by observers, who will then be able to extract astrophysical information from their observations quite efficiently, rather than only publish polarization curves, narrow band index curves, etc., with schematic analyses. It will be necessary to carry the idea through to completion with self-consistent solution of the whole problem, including an adjustment algorithm for **simultaneous** fitting of the several kinds of observations. Existing differential corrections and simplex programs for light curves or light and radial velocity curves can be generalized to handle n kinds of curves. Generalization of the WD program to include polarization curves will be accomplished fairly soon although, allowing for testing, a few years may pass before it is generally available. The worth of impersonal solutions will be appreciated by anyone who has tracked down several published values of a crucial astrophysical quantity, with no way to judge which one to believe or to know what are the uncertainties, because all were estimated subjectively. Simultaneous least squares fitting is particularly valuable because it not only eliminates guesswork about whom to believe, but provides a logical basis for deciding what to believe.

Considerable work is needed to create programs to do simultaneous solutions of light, velocity, polarization, narrow band index, ..., etc. curves, but it is just work – the principles have all been established. Weighting schemes appropriate to simultaneous multicolor light and velocity curve solutions are discussed in Wilson, 1979, and the extensions to other types of curves are obvious. The advantages to be realized by adding polarization and narrow band indices are much greater than those of simultaneous light and velocity solutions because more parameters are in common. With light and velocity curves, effectively only a few parameters are in common (rarely more than two and sometimes only one). Although the present simultaneous solution program (Wilson, 1979) is for the method of differential corrections, there is no reason not to apply the idea to simplex programs also. The power of desktop workstations now gives us the exhilirating prospect of finding one coherent solution which self-consistently satisfies properly weighted light, radial velocity, polarization, and other curves.

The entire idea of learning about stars from observations of close binaries is to make use of observational quantities with the following characteristics:

1. We understand how the quantities are formed (i.e., how each quantity can be synthesized from a realistic model).
2. The quantities can be accurately and reproducibly measured.

3. The quantities are sensitive to model properties.

In addition, for most quantities we prefer an intricate but reproducible variation in time, and we prefer information which does not duplicate that from other sources. Light curves and radial velocity curves are classic examples of observations which satisfy those requirements. There now are others through which, aided by modern computing power, we can achieve insight in new ways.

Acknowledgments

I thank J. C. Liou, J. D. Mukerjee, G. J. Peters and D. Terrell for providing previews of their research, which is soon to be published, for this overview paper. I also thank G. D. Henson for sending polarization curves of Algol, and E. F. Guinan for Figure 2. I am pleased to thank E. F. Milone for the invitation to present this paper at the 1991 I.A.U. General Assembly. The work was supported under the ISRP program of the Florida Space Grant Symposium (M. A. Eisenberg Director – NASA Grant NGT-40015). Financial support to attend the General Assembly was provided by the American Astronomical Society, and by the Division of Sponsored Research and the Astronomy Department of the University of Florida.

References

Bell, R. A. 1971, MNRAS, 154, 343

Brown, J. C., McLean, I. S., and Emslie, A. G. 1978, A & A, 68, 415

Chandrasekhar, S. 1946, ApJ, 103, 351

Code, A. D. 1950, ApJ, 112, 22

Collins, G. W. 1970, ApJ, 159, 583

Collins, G. W. 1972, ApJ, 175, 147

Corcoran, M. F., Siah, M. J., and Guinan, E. F. 1991, AJ, 101, 1828

Crawford, D. L. 1973, in "Problems of Calibration of Absolute Magnitudes and Temperatures of Stars", ed. B. Hauck and B. E. Westerlund (Dordrecht: D. Reidel Publ. Co.), p. 93

Crawford, D. L. 1975, AJ, 80, 955

Drake, S. A. and Ulrich, R. K. 1980, ApJ Suppl, 42, 351
Eaton, J. A., Hall, D. S. and Honeycutt, R. K. 1991, ApJ, 376, 289
Gies, D. R. and Wiggs, M. S. 1991, ApJ, 375, 321
Harrington, J. P. 1969, Ap Letters, 3, 16
Kemp, J. C., Henson, G. D., Barbour, M. S., Kraus, D. J., and Collins, G. W. 1983, ApJ Lett, 273, L85
Krolik, J. H. and McKee, C. F. 1978, ApJ Suppl, 37, 459
Lin, D. N. C. and Pringle, J. E. 1976, in "Structure and Evolution of Close Binary Systems", ed. P. Eggleton (Dordrecht: D. Reidel Publ. Co.), p. 237
Lubow, S. H. and Shu, F. H. 1975, ApJ, 198, 383
Mukherjee, J. D., Peters, G. J., and Wilson, R. E. 1992 (this volume).
Prendergast, K. H. and Taam, R. E. 1974, ApJ, 189, 125
Richards, M. T. 1992, ApJ (preprint)
Scharlemann, E. 1981, ApJ, 246, 292
Stromgren, B. 1966, Ann. Rev. Astron. Ap., 4, 433
Tassoul, J. L. 1987, ApJ, 322, 856
Tassoul, J. L. and Tassoul, M. 1990, ApJ, 359, 155
Terrell, D. and Wilson, R. E. 1992 (this volume).
Whitehurst, R. 1988a, MNRAS, 232, 35
Whitehurst, R. 1988b, MNRAS, 233, 529
Wilson, R. E. 1979, ApJ, 234, 1054
Wilson, R. E. 1990, ApJ, 356, 613
Wilson, R. E. 1991, in "Pacific Rim Colloquium on New Frontiers in Binary Star Research", ed. K. C. Leung, (Astron. Soc. Pacific Conference Series, in press).
Wilson, R. E. and Liou, J. C. 1992 (to be published).
Zahn, J. P. 1977, A & A, 57, 383

Discussion

Q. (J. Kallrath to R. E. Wilson): How many additional parameters are needed for polarization and narrow bandwidths?

A: About 5 to 10, depending on the sophistication of the model.

Q. (S. Soderhjelm to R. E. Wilson): Is the sense of revolution for the close pair in Algol still opposite to the wide orbit? If so, I am a bit worried.

A: It is still in the opposite sense. There are some known counter-rotating binaries. Of course, counter-rotation of the projected motions is not the same as counter-rotation in the 3-dimensional sense.

Q. (Andrej Cadez): Are your binary system computer programs available?

A: Yes, the programs are, or soon will be, available to those who wish to use them. The latest versions (not incorporating polarization) are currently being debugged, and should be available in approximately 1 year. The programs which include all the parameters which have been discussed here may be available in approximately 4 years.

Spectral Energy Distributions of Circumstellar Gas in Algols

Dirk Terrell and R.E. Wilson
Astronomy Department, University of Florida, Gainesville, FL 32611, USA

I. Introduction

Developments in computers now have made it practical to treat the gas flow and radiation problem in matter-transferring binaries. The computation consists of four separable parts: (1) gas flow, (2) radiative properties, in particular, emission line strengths and profiles, (3) generation of spectral energy distributions by spatial integration with Doppler shifts and eclipses, and (4) adjustment of model parameters based on comparisons with observational data. Our hydrodynamical code is a 3–dimensional Lagrangian scheme which eliminates the memory and computing limitations of finite-difference methods, and does not constrain the global geometry of the system artificially. The gravitational acceleration of the gas is computed via Newtonian mechanics. Viscous and pressure effects are included and can be followed continuously in time over many cycles. Viscous interaction among gas particles is effected by allowing neighboring particles to exchange momentum. Pressure gradients are computed by smoothed particle hydrodynamics.

Emission line strength and its time variation is an important output of the overall computation. The emission line strength needs to be modeled with a computational module which can readily be replaced by improved modules. An important simplification is that self irradiation by the gas can be neglected, at least for now. Another important element of the output is the computation and spatial integration of Doppler shifts. A major simplification in this area is that intrinsic line profiles are unimportant compared with orbital Doppler broadening. The line strength and Doppler information appear together in the output in the form of a wavelength-dependent flux, but can be regarded separately from a computational point of view.

We are developing a modular program which is the analog (with respect to circumstellar gas) of a binary star light curve program. The modular nature of the program increases its flexibility because computations are done in subroutines which can easily be modified or replaced as better techniques are developed. Confrontation between theory and observation will test the usefulness of our model. At present, the major source of observations is the archival collection of data from the International Ultraviolet Explorer, although there is some useful ground-based optical and infrared spectrometry. We hope that the existence of the model will stimulate spectrometry with more nearly continuous coverage in time.

II. Hydrodynamics

A. Previous work

The existence of gas streams in Algols has been known for about fifty years (Joy, 1942; Struve, 1944). Early attempts to compute the flow of the gas were solutions of the restricted three-body problem for particles ejected from the inner Lagrangian point at arbitrary angles and speeds (Gould, 1957, 1959; Plavec, Sehnal, and Mikuláš, 1964). These studies demonstrated the gross properties of the flow, but Prendergast (1960) pointed out that the proper treatment of the problem is a hydrodynamical one since the mean free path of the gas particles is much smaller than the size of the flow region. Because of computing limitations, however, a full hydrodynamical treatment was not achieved until Prendergast and Taam (1974) computed the steady-state flow for several cases. Their method is based on the fact that the equations of hydrodynamics are moment equations of the Boltzmann equation. Rather than solve a set of difference equations, they simulate a solution of the Boltzmann equation. A drawback to their method is that the viscosity is implicitly tied to the size of a grid box, and thus one cannot alter the viscosity without changing the resolution of the computation.

Lubow and Shu (1975) made use of matched asymptotic expansions and showed that particle-trajectory calculations can be useful over a large region which they call the "orbit region," since the mean free path argument is relevant only if particle trajectories try to cross one another. They also showed that material leaves the L_1 point in a narrow stream at an angle mildly dependent on the mass ratio of the system (see their Table 1). In a later paper (Lubow and Shu, 1976) they discussed the vertical structure of the gas stream.

Lin and Pringle (1976) used a modified particle approach to follow the gas flow in the orbital plane. Their method allows for viscous interaction between particles by allowing them to share momentum in a dissipative process, but ignores pressure gradients. To effect viscosity, they allocate the particles to the boxes of a grid centered on the mass-gaining star, and then allow particles within each box to interact with one another. Unfortunately, this scheme has the undesirable consequence that particles that are close to one another, but not in the same box, will not interact, while particles that are more remote *will* interact simply because they are in the same box. We present a modification to this scheme that eliminates this problem.

Recently, Whitehurst (1988a, 1988b) has developed a code based on the Lin and Pringle method, but modified to include viscous and pressure forces in the manner described by Larson (1978). Larson's method regards the particles as extended, deformable gas clouds which are in continual contact. The repulsive acceleration between particles is given by $a_r = \frac{C^2}{r}$ where C is a parameter proportional to the local sound speed and r is the separation of the particles. Whitehurst applied his model to systems with degenerate primaries (gainers).

B. Our program

In order to avoid the memory and computing speed limitations of finite difference schemes, we have adopted a 3–dimensional Lagrangian scheme similar to the 2–D method used by Whitehurst. The basic assumptions about the system are that the stars act as centrally condensed mass points and move in circular orbits. The coordinate system employed is one whose origin is at the center of mass of the system, with the unit of length being the separation of the two stars. The x-axis of the system includes the center of mass of each star and is positive in the direction of the mass-losing star. The z-axis is the axis of rotation and the y-axis is defined so that the system is right-handed. The unit of time is $P/2\pi$, where P is the period of rotation of the system. We define the mass ratio as that of loser to gainer, *i.e.*

$$q = \frac{m_2}{m_1}$$

It is convenient to use the auxiliary quantity

$$\mu = \frac{q}{1+q}$$

and write the acceleration components of a particle at (x,y,z) due to gravitational forces as

$$\ddot{x} = 2\dot{y} + x - \frac{1-\mu}{r_1^3}(x - x_1) - \frac{\mu}{r_2^3}(x - x_2)$$

$$\ddot{y} = -2\dot{x} + y - \frac{1-\mu}{r_1^3} - \frac{\mu}{r_2^3}y$$

$$\ddot{z} = -\frac{1-\mu}{r_1^3}z - \frac{\mu}{r_2^3}z$$

where

$$r_1^2 = (x - x_1)^2 + y^2 + z^2$$

$$r_2^2 = (x - x_2)^2 + y^2 + z^2$$

with x_1 and x_2 being the distances of the primary and secondary from the center of mass of the system.

The user can select how the particles are ejected from L_1 by specifying two angles— ϕ, an angle measured counterclockwise from the positive x-axis and in the plane of the orbit and ψ, an angle measured from the positive z-axis. For example, ϕ would be 180° for ejection toward the primary component. A value of ψ = 90° will give ejections in the plane of the orbit. The program also allows for ranges of these angles, $\Delta\phi$ and $\Delta\psi$. The ranges are centered on the specified angles and particles are emitted randomly over the ranges. The user also specifies the (dimensionless) ejection speed. In our experiments, particles are ejected from L_1 with initial conditions consistent with the calculations of Lubow and Shu (1975).

The viscosity is simulated by allowing particles within a specified distance, σ, of one another to share momentum. At present, we use particles whose masses are equal, and thus the mass contribution to the momentum may be neglected. If two particles have a separation $r < \sigma$, and are approaching one another, they are allowed to interact. The interaction is such that the particles retain their positions, but have their velocity components modified. The new velocity components in the x-direction for each particle are computed as follows:

$$\dot{x}_{avg} = \frac{(\dot{x}_1 + \dot{x}_2)}{2}$$

$$\dot{x}_1(new) = \dot{x}_{avg}\left(1 - \frac{r}{\sigma}\right) + \dot{x}_1\left(\frac{r}{\sigma}\right)$$

and

$$\dot{x}_2(new) = \dot{x}_{avg}\left(1 - \frac{r}{\sigma}\right) + \dot{x}_2\left(\frac{r}{\sigma}\right)$$

The y and z-components are, of course, done in the same way. This procedure allows the user to control the viscosity through the variable σ.

Pressure gradients are computed by the method of smoothed particle hydrodynamics (SPH) as developed by Lucy (1977) and Gingold and Monaghan (1977, 1978, 1979a, and 1979b). Basically, SPH treats fluids elements as extended clouds of material whose centers of mass move according to the conservation laws of hydrodynamics. To compute values of continuous variables such as density and pressure from the discrete distribution of particles, an average over the particle distribution is taken by using an appropriate "smoothing function." This procedure has the advantage that spatial derivatives of the continuous quantities can be replaced by the analytically calculated derivatives of the smoothing function.

The smoothing function, $w(\mathbf{r}-\mathbf{r}')$, must satisfy

$$\int_V w(\mathbf{r}-\mathbf{r}_j)d^3\mathbf{r} = 1$$

where V represents the volume of interest and $\mathbf{r}_j$ is the position vector of particle j. With n particles of mass m, the smoothed density is given by

$$\varrho(\mathbf{r}) = m\sum_{j=1}^{n} w(\mathbf{r}-\mathbf{r}_j)$$

and the acceleration acting on each particle due to pressure gradients is

$$\frac{d\mathbf{v}_i}{dt} = -m\sum_j\left(\frac{P_i}{\varrho_i^2}+\frac{P_j}{\varrho_j^2}\right)\nabla w(\mathbf{r}-\mathbf{r}_j)$$

Schüssler and Schmitt (1981) point out the necessity of carefully selecting the form of the smoothing function. They recommend a smoothing function of the form

$$w(r) = \{\begin{matrix} c\,[1-\sqrt{\frac{r}{h}}\,] & for\ r \le h \\ 0 & for\ r > h \end{matrix}$$

where the normalization constant is given by

$$c = \frac{21}{4\pi h^3}$$

and h is the user-specified smoothing length. This smoothing function avoids artificial clustering that occurs with other smoothing functions in situations where self-gravity is neglected, and we have adopted it in our code.

The viscosity and pressure calculations are done in separate subroutines, and control integers in the input file can be used to include or not include these calculations. Of course, with the control integers both set to skip viscosity and pressure calculations, the simulation becomes a restricted three-body trajectory calculation. At each time step, the particles are allowed to interact with one another, and the pressure and gravitational forces are computed. The particle positions are then integrated with a fourth order Runge-Kutta technique.

A potentially very time-consuming part of the computation is determining which particles should interact with one another. An inefficient way of achieving this would be to compute the separations of all pairs of particles. Obviously, this would require large amounts of time and memory. A much more efficient method would be to assign the particles to the boxes of a grid, where the box size is at least as large as the interaction length. Then, to determine which particles will interact with a certain particle, one need only search over the box containing the particle and the eight boxes surrounding it (Since the disks tend to be of small vertical extent, a 3–dimensional grid is not necessary.). This scheme eliminates the problems discussed previously for the Lin and Pringle (1976) scheme, and the (artificial) grid serves only as a computational aid, not affecting the particles in any way. Since there are two interaction lengths (viscosity and pressure), there should be two grids. In our present program, however, we employ only one grid, with the box size being the larger of the interaction lengths. Since the two lengths are usually similar, using only one grid does not impose a large penalty in computing efficiency. Figure 1 shows a developing flow pattern.

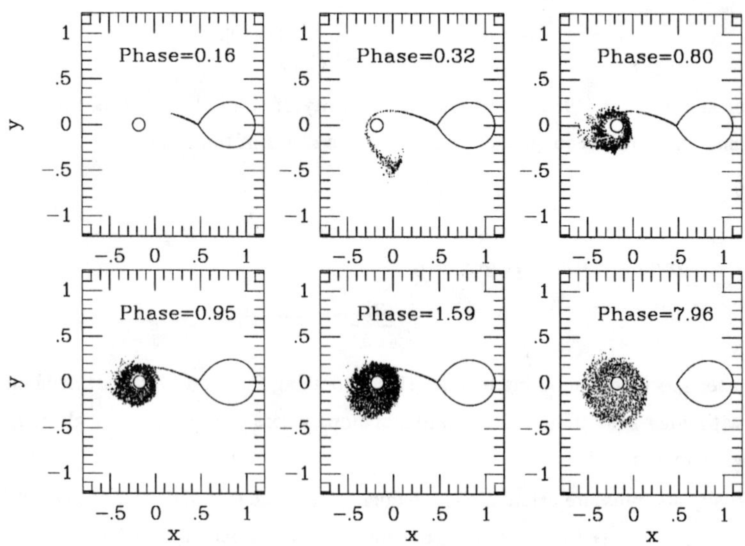

Figure 1 Episodic mass flow event in SX Cas and subsequent disk formation at various phases.

III. Radiation

Previous simulations of the mass transfer in Algol-type systems have made, at best, only crude attempts to predict observational consequences of the radiation that arises from the gas. The reason for this is that computing either the gas flow or its radiative properties is itself quite a computational endeavor. To do both parts of the problem has, until only recently, been computationally possible only on supercomputers. To simulate the fully coupled radiative transfer and hydrodynamics problem is still a formidable task. Fortunately, one may apply certain simplifying assumptions to the radiative transfer problem.

At present, we are in the early stages of the development of the radiation portion of the code. For the moment, we consider only hydrogen, although eventually we will include other important species. The gas is assumed to be completely ionized and to behave as an ideal gas. We also assume that optical depths in the gas are small enough that self-irradiation may be neglected.

The radiative properties of a cloud of hydrogen under varying conditions have been studied by many groups. The classical theoretical approach taken by Menzel and Baker (1938) assumed that the Lyman lines were optically thick or completely optically thin. Less restrictive assumptions were employed by Krolick and McKee (1978) and Drake and Ulrich (1980, hereafter DU) in employing the escape probability approach. The DU code is more appropriate for Algols, because the method is valid for electron densities as high as 10^{15} cm^{-3}. We will not attempt to give the details of their approach, except to say that it includes all relevant collisional processes, explicitly treats the angular momentum (l) sublevels for the first six energy (n) sublevels, assumes statistical equilibrium of the l sublevels for $n>6$, treats energy levels explicitly for $n\leq30$, omits stimulated emission terms due to the external radiation field, and omits excited-state photoionizations. DU give justifications for these assumptions which are applicable to the Algol regime.

To compute line profiles, the density, temperature, and velocity of the gas at each particle location are taken from the hydro portion of the code. The first calculation is to check whether the gas element is eclipsed by either star. Presently, we assume that both stars are spherical (the secondary being of radius r_{side}, *i.e.*, the relative radius of the Roche lobe in the plane of the orbit perpendicular to the line of centers). If the gas element is not eclipsed, the temperature and density are then used to compute the line profile from a static parcel of gas at those conditions. The radiation is then Doppler shifted and allocated to bins of specified width (in our runs, 0.1 Å). The program sums the flux contributions of all particles. Figure 2 shows Hα profiles as seen from various aspects for a disk in SX Cas as it appears 1.59 binary cycles after a mass transfer event

Eventually, this program will be interfaced with the Wilson-Devinney light curve program (Wilson and Devinney, 1971; Wilson, 1979; Wilson, 1990) to compute the radiation from the two stars. This will remove our present simplistic assumptions about the figures and radiation fields of the stars and will enable us to model light curves "disturbed" by contributions from the gas, such as the light curve of U Cep (*e.g.*, Markworth, 1979).

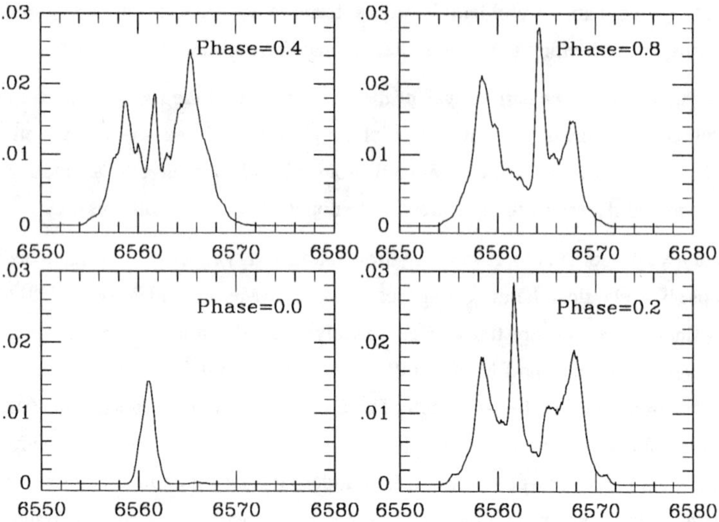

Figure 2 The appearance of the Hα line seen at various phases for a disk formed 1.59 cycles after a mass transfer event in SX Cas. The vertical scale is in arbitrary flux units and the horizontal scale is in Ångstroms.

IV. Comparison with Observations

This stage of the project is the least developed. Since we have only recently achieved the synthesis of line profiles, only a qualitative comparison with observations has been made. However, we are encouraged by the results of such comparisons. Kaitchuck (1990) has argued convincingly from his observations of the disk in RY Gem, that it is significantly asymmetric. Although we have not yet tried to simulate the disk in RY Gem, our results for SX Cas have similar properties.

One extensive source of data for these systems is the archival data base of the International Ultraviolet Explorer (IUE) satellite. Since most of the emission arising from the disk is in the ultraviolet, space-based observations will likely be the primary source of data. Another factor in favor of space observatories is their ability to monitor objects practically continuously, as they are unaffected by clouds and the diurnal cycle. Because these mass transfer events are episodic, it is necessary to follow them continuously over several binary orbital cycles. Such observations have not typically been done in the past, but we hope that they will be stimulated by the existence of our model.

The first comparisons with observations will likely be subjective. With experience, we will develop impersonal parameter adjustment algorithms. This will finally allow quantitative comparisons with observations to better understand the specifics of mass transfer and binary star evolution.

V. Acknowledgements

D.T. would like to acknowledge the generous assistance of Steve Drake in supplying his radiation code and providing help in running it. D.T. is supported by a NASA Graduate Student Researcher Fellowship. Travel of both authors to the meeting was supported in part by the American Astronomical Society's International Travel Grant Program. R.E.W. received additional travel support from the Division of Sponsored Research and the Astronomy Department of the University of Florida.

VI. References

Drake, S.A. and Ulrich, R.K. (1980), Ap. J. Sup. **42**, 351.

Gingold, R.A. and Monaghan, J.J. (1977), M.N.R.A.S. **181**, 375

Gingold, R.A. and Monaghan, J.J. (1978), M.N.R.A.S. **184**, 481.

Gingold, R.A. and Monaghan, J.J. (1979a), M.N.R.A.S. **188**, 39.

Gingold, R.A. and Monaghan, J.J. (1979b), M.N.R.A.S. **188**, 45.

Gould, N.L. (1957), P.A.S.P. **69**, 541.

Gould, N.L. (1959), A.J. **64**, 136.

Joy, A.H. (1942), P.A.S.P. **54**, 35.

Kaitchuck, R (1990), P.A.S.P. **100**, 594.

Krolick, J.H. and McKee, C.F. (1978), Ap. J. Sup. **37**, 459.

Larson, R.B. (1978), J. Comp. Phys. **27**, 397.

Lin, D.N.C. and Pringle, J.E. (1976), In: *IAU Symposium No. 73, Structure and Evolution of Close Binary Systems*, eds. Eggleton, P. *et al.*, pp. 237–252, Reidel, Dordrecht, Holland.

Lubow, S.H. and Shu, F.H. (1975), Ap. J. **198**, 383.

Lubow, S.H. and Shu, F.H. (1976), Ap. J. **207**, L53.

Lucy, L.B. (1977), A.J. **82**, 1013.

Markworth, N.L. (1979), M.N.R.A.S. **187**, 699.

Plavec, M., Sehnal, L., and Mikuláš (1964), B.A.C. **15**, 171.

Prendergast, K.H. (1960), Ap. J. **183**, 162.

Prendergast, K.H. and Taam, R.E. (1974), Ap. J. **189**, 125.

Schüssler, M. and Schmitt, D. (1981), Astr. Ap. **97**, 373.

Struve, O. (1944), Ap. J. **99**, 222.

Whitehurst, R. (1988a), M.N.R.A.S. **232**, 35.

Whitehurst, R. (1988b), M.N.R.A.S. **233**, 529.

Wilson, R.E. (1979), Ap. J. **234**, 1054.

Wilson, R.E. (1990), Ap. J. **356**, 613.

Wilson, R.E. and Devinney, E.J. (1971), Ap. J. **166**, 605.

VII. Discussion

Q. (A. H. Batten to D. Terrell): How do your model calculations compare to Kaitchuck's observations?

A: Detailed comparisons have not been made but the computed disk has the properties that Kaitchuck has observed (i.e., strongly non-Keplerian motion). Very soon we will be making more detailed comparisons with his work and that of others and I suspect that we will find good agreement.

Q. (J. Kallrath to D. Terrell): Can you estimate the effect of shocks when the stream hits the disc?

A: We are concentrating on that now. In particular for SX Aurigae, shocks are not very strong, and that is why we chose it for these models.

Q. (K.-C. Leung to D. Terrell): Have you considered shock phenomena?

A: Not in these simulations, but we will be including them soon.

Q. (Bilhu to D. Terrell): What kind of viscosity is being used?

A: We simulate the viscosity in a new way. Particles within a specified distance of one another are allowed to share momentum. The velocity components of the particles are averaged. The particles are then given a fraction of the average velocity component, based on the separation. Closer particles are given a larger fraction than more distant ones. Also, I failed to say that only particles that approach one another can interact.

Q. (Zwitter to D. Terrell): Does inclusion of the 3rd dimension produce differences from 2- dimensional models?

A: The models here were at a mass transfer rate that was small enough to cause the disk to rapidly settle into the orbit plane. Other experiments have shown that a higher transfer rate leads to a disk which retains a vertical structure for a longer time, but we have not yet made a quantitative assessment of that structure.

Q. (G.A. Richter to D. Terrell): I think your method will probably be suitable for the examination of superhump theories in SU UMa systems. Do you agree?

A: Yes. So far we have concentrated on systems with non-degenerate primaries. There may be other important processes to model in the case of a degenerate primary, like magnetic fields for example, but the flexibility of the program will make the incorporation of those processes quite simple.

Gradient Free Determination of Eclipsing Binary Light Curve Parameters - Derivation of Spot Parameters Using the Simplex Algorithm

Josef Kallrath
BASF-AG, ZXT/C, Kaiser-Wilhelm-Str. 52,
D-6700 Ludwigshafen, Germany
and
Astronomische Institute der Universität Bonn, Auf dem Hügel 71,
D-5300 Bonn 1, Germany

Abstract. During the last twenty years many eclipsing binary light curves have been analyzed by using theoretical light curve models and the method of *differential corrections* for parameter estimation. This method requires an initial guess not too far away from the best fit parameter set. The disadvantage of local convergence properties can be overcome by the use of gradient free optimization procedures. A comparison of both approaches is given, and a summary of experiences on the use of the *simplex algorithm* in connection with the Wilson-Devinney model is reported. Finally, a Kurucz atmosphere, simplex-enhanced version of the 1983 Wilson-Devinney model, which has been extended and now allows for the determination of spot parameters also, is presented.

1 Introduction

The intention of this work is to give a brief overview on eclipsing binary light curve optimization techniques, to review experiences with the simplex algorithm, and to present an extension to the Wilson-Devinney program which allows for the estimation of an arbitrary number of spot parameters.

The analysis of photometric and spectroscopic data of eclipsing binaries has been performed during the last twenty years by means of synthetic light curves from which the parameters like mass ratio q, inclination i, temperature T_2 and other quantities have been derived with some algorithms capable of solving non-linear least squares problems. The procedures in use differ significantly with respect to the physical model [spherical stars in EBOP (Eclipsing Binary Orbit Program, Nelson & Davis 1972, Etzel 1981), ellipsoidal geometry in WINK (Wood 1971), and Roche geometry in the Wilson-Devinney program (Wilson & Devinney 1971,Wilson 1979)], and also in the method used to solve the non-linear least squares problem : *differential corrections* (Wilson *et al.* 1972), Napier

1981, *simplex algorithm* (Kallrath & Linnell 1987)]. Thus, the results, i.e. the physical parameters derived from the photometric data are depending on the model, and to some extent on the least-squares solver also. In principal, the results should be the same for any least-squares solver. Implicitly, we assume that this is true for the rest of this paper, and we concentrate on those differences as convergence properties and computational effectiveness and efficiency which are most important for all practical purposes. The physical models in discussion are those developed by Wilson & Devinney (1971) and later versions of the same authors. For a brief review on these models (and the Nelson-Davis-Etzel model also) we refer to Etzel & Leung (1990). Actually, we would like to distinguish between all the versions published between 1971 and 1978 (referred to as WD1), and all newer versions (Wilson 1979) between 1979 and 1990 (referred to as WD2). The Wilson-Devinney model has been chosen for its increased popularity: In the 1988 IAU Commission's 42 report (Smak 1988) it has the highest percentage of light curves studied by different light synthesis procedures.

Thus, the goals and the structure of this paper is as follows : In Section 2 the mathematical structure of the non-linear least-squares problem is briefly reviewed. In Section 3 some disadvantages of the method of *differential corrections* with respect to its local convergence properties are listed and it is described how they can be overcome by the use of gradient free optimization procedures also in use since many years. A comparison of both approaches is given, and a summary of experience gained from the use of the simplex algorithm in connection with the Wilson-Devinney model is reported. Finally, in Section 4 we present a Kurucz atmosphere, simplex-enhanced version of the 1983 Wilson-Devinney model which has been extended and now allows for the determination of spot parameters also.

2 The light curve analysis

The usual approach to light curve synthesis is the following: Given a physical model and a parameter vector $\mathbf{x} = (x_1,x_2,\ldots,x_m)$, we seek a solution in multidimensional parameter space minimizing the quadratic form $(\mathbf{d},\mathbf{Wd})$, where $\mathbf{d} = \mathbf{o}-\mathbf{c}$ is the vector of light residuals (observed minus calculated intensities) and $W_{i\nu} = w_i \delta_{i\nu}$ specifies the assigned individual weights of the observations. As a measure for the quality of fit we take

$$\sigma_{fit} := f(\mathbf{x}) = \{ (\frac{n}{n-m} / \sum_{\nu=1}^{n} w_\nu) \cdot \sum_{\nu=1}^{n} w_\nu \cdot [d_\nu(\mathbf{x})]^2 \}^{1/2} , \qquad (2.1)$$

since this quantity is normalized with respect to the number of observed data points and number of adjusted parameters. In this form, we have a typical situation for fitting model functions (here: the light curve model) to data, i.e. nonlinear parameter estimation, which in this form is a special case of unconstrained optimization. However, strictly speaking, we are confronted with a nonlinear *constrained* optimization problem since the optimization is subject to bounds

$$l_i \leq x_i \leq u_i \ . \qquad (2.2)$$

Typical boundaries are $l_i=0$ for most of all parameters x_i, and $u_i=1$ for albedos, limb darkening coefficients, or gravity darkening coefficients. Sometimes some limits on the inclination i or the mass ratio q are known in advance. The optimality conditions, i.e. necessary and sufficient conditions for the existence of a minimal point $\mathbf{x}$ of (2.1,2.2) can be expressed in terms of the first and second order derivatives and some appropriately chosen Lagrange multipliers. The non-linear structure of the problem requires an iterative algorithm for solving the problem.

Another problem, which may occur, is the existence of different local minima. Physically, in eclipsing binary light curve analysis, their existence is justified by possibility of transit and occultation minima. Both assumptions may lead to solutions with σ_{fit} values of comparable size as experienced with BF Aurigae in Kallrath & Kämper (1992). Such non-uniqueness problems may be overcome by some additional information, e.g. line ratios in a two-spectra system.

For a further discussion on the requirements and limits of the use of synthetic light curves for parameter determination we refer to Etzel & Leung (1990).

3 Differential corrections and alternative algorithms

3.1 Some characteristics of the method of differential corrections

In the past many analyses of photometric light curve data of close binaries were performed with the Wilson-Devinney program. Wilson and Devinney (1971) used the classical method of *differential corrections* to optimize light curve solutions. This method belongs to a class of Newton-type methods without second derivatives. It makes use of the fact that the gradient and the Hessian matrix of $f(\mathbf{x})$ have a special structure, but is based on the premise that finally first-order terms will dominate second-order terms which is not justified when the residuals at the solution are very large. In practice it turns out that this assumption seems to be valid, i.e. the residuals at the solutions are small enough. The bounds (2.2) are not taken into account by the method of *differential corrections*. The procedure has been applied successfully if an initial solution $\mathbf{x}_0$ close enough to the final solution $\mathbf{x}^*$ is known and if the correlations between the parameters are not too large.

However, there are some often encountered problems with *differential corrections:*

1. It is based on local convergence properties and one may easily run into problems of divergence, if the initial approximation is not good enough.
2. Even when the initial approximation is close to a minimum on the WSSR hypersurface, the effects of nonlinearity may be detrimental to an optimization via differential corrections
3. Differential corrections tends to oscillate, i.e. σ_{fit} is not a monotonically decreasing function of the number of iterations (a step length control might help).
4. the normal equations may become ill-conditioned when using a larger number of free parameters (complete orthogonal factorization or singular value decomposition is recommended).
5. The partial derivatives have to be calculated numerically. This means that to adjust m parameters, m+1 or 2m+1 light curves have to be calculated at each iteration, depending on whether asymmetric or symmetric differences are used in calculating partial derivatives. Their accuracy depends critically on the choice of the increments and is difficult to control. In addition, the requirements on precision, hence grid density for the calculation of the theoretical light curve by numerical quadrature of the flux over the star's surface are obviously much higher than for a direct search method. In the past, the high computational cost have prevented many light curve analyzers from using all observed data point. Instead, they formed *normal points* and used them to estimate the parameters. We do not recommend this procedure at all since systematic effects changing the solution are difficult to control.
6. It is very difficult to investigate the uniqueness of a solution. In particular, those questions become important when analyzing light curves with primary and secondary minimum of similar depths. In that case, e.g. *BF Aurigae* (Kallrath & Kämper 1992), there occur two well-separated local minima for σ_{fit}.

3.2 Differential corrections versus simplex algorithm

These well-known disadvantages of *differential corrections,* especially its inability to perform a global search in parameter space and the large CPU time needed, have led to the application of the simplex algorithm (Spendley *et al.* 1962, Nelder & Mead 1965) to the solution of light curves as described in Kallrath & Linnell (1987). Its basic idea is to compare the standard deviation at the vertices of a simplex moving through parameter space by some simple geometric operations. Those parameter combinations with the highest value are eliminated consecutively while the simplex is contracting towards the final minimum on the σ_{fit}-hypersurface. For further mathematical background we refer to Kallrath & Linnell (1987) and the references cited therein. This global search method is computationally efficient, does not depend on derivatives, cannot diverge, and is capable of automatic iteration.

From our numerical experiments in connection with the analysis of ***MR Cygni*** (Kallrath & Linnell 1987, Linnell & Kallrath 1987), ***TU Muscae*** (Kallrath 1987) and ***V 781 Tauri*** (Schnell *et al.* 1992) we find the simplex algorithm very useful to perform the initial search in parameter space. Due to the computational efficiency of the *simplex algorithm* and its operational stability it is much easier to investigate questions of uniqueness. Therefore, it is of great help when investigating the whole parameter space and discussing problems of uniqueness as in the binary systems ***V 836 Cygni*** (Breinhorst *et al.* 1989) or ***BF Aurigae*** (Kallrath & Kämper 1992). In particular, those questions become important when analyzing light curves with primary and secondary minimum of similar depths. In the case of BF Aurigae we were able to decide on the question whether the primary minimum was a transit or occultation minimum. Being very close to the solution, the *differential corrections* converge more rapidly since it can make use of gradient information, so it may be efficient to switch from *simplex* to *differential corrections* in the late stages of the iteration process. Besides that, the correlation matrix which becomes available when using *differential corrections* indicates parameters which are only weakly determined.

Finally, one has to note that due to the non-linear structure of the problem both methods cannot guarantee the uniqueness of a *solution*, i.e. one is always forced to assume that the (local) minimum found is already the global one.

3.3 On experiences with the simplex algorithm in light curve analysis

The first application of the *simplex algorithm* to light curve analysis was performed and tested in Kallrath & Linnell (1987) when re-investigating ***MR Cygni***. This analysis was performed with LCCTRL (based on WD1), while, simultaneously, the algorithm was used to analyze ***MR Cygni*** on the background of Linnell's light curve program SYNPGM (Linnell 1984, Linnell & Kallrath 1987). This second installation made it obvious that the *simplex algorithm* can be installed within reasonable time into existing light curve code.

A preliminary analysis of ***BF Aurigae*** (Kallrath 1987), and then a detailed investigation (Kallrath & Kämper 1992) of this interesting binary system has shown that, supported by a somewhat changed program logic, it is possible to do the analysis without assumptions about the geometry of the system. There is a continuous transition between detached (mode 2), semi-detached (mode 4,5) and contact-system (mode 1,3). Thus it is sufficient to start with the most general mode, i.e. with a detached system, and then let the program find the most appropriate geometry.

To test the software in more details, in Kallrath (1987) LCCTRL as well as SYNPGM, and the *simplex algorithm*, were applied to the early spectral type (O7V+BOV) contact binary *TU Muscae*. In contrast to the analysis of Wilson & Rafert (1981), this investigation considered all 2800 data points of the superb *uvby* photometry of Andersen & Gronbech (1975). Furthermore, the interpretation of the system was supported with excellent spectroscopic data from the same authors. As expected from the quality of the data there were no convergence problems nor questions of uniqueness left open in this analysis.

Another near contact hot binary, *SX Aurigae,* has been investigated by Linnell *et al.* (1988) with SYNPGM and special focus on the determination of the bolometric albedo A_2. The uncxpected value A_2=0.4 was partly explained to be caused by the rapid rotation in early type stars.

Quite different as *TU Muscae* was the case of the EB-type short-period eclipsing binary *V 836 Cygni* (Breinhorst *et al.* 1989). The global minimum of σ_{fit} was only marginally pronounced, and thereby allowed the light curve to be fitted over a whole range of parameters (with a mass-ratio interval $0.3 \leq q \leq 0.5$). However, this analysis has shown how valuable a parameter study (grid solution along q axis) can be for deriving a consistent and detailed model of a binary system.

Drechsel *et al.* (1989) in their analysis of *LY Aurigae* and *AH Cephei* found the *simplex algorithm* especially useful for the quantitative determination of the third light contribution.

As an example for a semi-detached binary system we mention the analysis of the eclipsing binary *ST Carinae* by Walker & Chambliss (1990) who also compared both optimization procedures and checked for third light in the system.

The analysis of the W UMa Binary *V 781 Tauri* (Schnell *et al.* 1991) was started almost from scratch without any previous knowledge on the system, and proved the software's ability to be of great help during the initial search.

Two of the three NGC 5466 cluster binaries (Kallrath *et al.* 1991) showed a clear O'Connell effect. One way to overcome this problem is to include spots in the light curve modeling. For *H 235* (Stagg *et al.* 1991) in the open cluster NGC 752 there were radial velocity and light curves available. Therefore, taking the Kurucz-atmospheres version (Milone *et al.* 1991b) of the 1983 Wilson-Devinney program (Wilson 1979, WD2), the *simplex algorithm* was implemented in this software. Now, the program can be used to support simultaneous fitting of radial velocity and light curve, and spot fitting also.

Finally, we would like to stress that all analyses mentioned above have been performed with all data points available in the light curves, i.e. no *normal points* were used.

3.4 On alternative approaches to the estimation of light curve parameters

We should mention that there have been other attempts to overcome some of the limitations of *differential corrections*: Wilson and Biermann (1976) suggested to reduce the number of free parameters and introduced the "*method of multiple subsets*", by which they separate the most correlated parameters during the optimization process (which still proceeds by differential corrections). Various semi-empirical modifications of *differential corrections* were outlined by Khaliullina and Khaliullin (1984) and implemented in a computer program capable of automatic iteration (which, however, is designed to analyze the light curves of eclipsing binaries with spherical components, only). Hill (1979), in his program LIGHT, which is based on Roche geometry, has changed the light curve solver from simple *differential corrections* to the more sophisticated *Marquardt algorithm* (Marquardt 1963), which has much better convergence properties and has become the standard of non-linear least-squares routines. We did not try the *Marquardt algorithm* with our program but it could be worth an attempt. However, it still makes use of partial derivatives and that was what we tried to avoid with respect to the limited computer facilities in the mid eighties. Actually, nowadays with substantially improved resources we recommend to use the *simplex algorithm* for initial search and global inve-stigation, and to switch to the modified Levenberg-Marquardt algorithm (Press *et al.* 1989) when approaching a local minimum. This procedure is implemented in the MicroWAY product MINSQ (MicroMath 1991), and combines efficiently a steepest descent method with a step-size controlled Gauss-Newton algorithm. A similar step in this direction has already been performed by Plewa (1988). In his MINW method he combines the simplex algorithm, a variable metric method (Fletcher 1970), and the computation of the Hessian matrix for statistical purposes and error estimation. Unfortunately, his analysis of the contact binary ADS 9019 B is still based on normal points.

In eclipsing light curve analysis there have also been in use alternative gradient free optimization techniques. Napier (1981) uses a sequential (*creeping random*) search technique for fitting the parameters in his triaxial ellipsoid-ellipsoid geometry light curve model. Since his light curve model did not require a large amount of computer time he was able to find a solution parameter set within minutes. However, a similar approach namely a 'Controlled Random Search optimization prodecure' based on the *Price algorithm* (Price 1976) was applied to the Wilson-devinney model by Barone *et al.* (1988,1990). In this case the efficiency was very low. Barone *et al.* report solutions which required about 30000 iterations and 11 days of CPU time on a VAX 8600. This is probably not the royal road to the estimation of light curve parameters.

4 Spot Modeling

4.1 An enhanced version of the Wilson-Devinney for spot fitting

The difficulties occurring in the analysis of eclipsing binary light curves which show an O'Connell effect (unequally high consecutive light curve maxima), might be overcome by including star spots in the model (Yamasaki 1982, Davidge & Milone 1984, Poe & Eaton 1985, Milone *et al.* 1987, Hill *et al.* 1990). For a more complete review on spot modeling we refer to Linnell (1992). Budding & Zeilik (1987) and Zeilik *et al.* (1988) use adjustable spot parameters to represent the light curves of short-period RS CVn stars. The first time, spots are adjusted within a Wilson-Devinney model is found in Kang & Wilson (1989). However, due the logic of the software, when Kang & Wilson do spot fitting they have to fix some of the other light curve parameters. Thus, a Wilson-Devinney version supporting simultaneous fits of spot and other light curve parameters is missing. To reach that goal, as a model and a basic piece of software which is enhanced by the simplex algorithm we use an improved version (eccentric orbits, simultaneous fitting of photometric and spectroscopic data, spot modeling) of the Wilson-Devinney program (Wilson 1979) namely that of Milone *et al.* (1991,1992). The major extension to be mentioned is the option to include the Kurucz atmosphere. This atmosphere option certainly improves the model since it provides a more realistic treatment of the radiation physics, and even considers the finite width of UBV photometry. Recently (Kallrath *et al.* 1992), this version has been further extended to allow both the *differential corrections* and the *simplex algorithm* for parameter fitting. With the *simplex algorithm* it is also possible to determine automatically the four spot parameters (latitude φ, longitude λ, radius ρ, and temperature factor t_f) of one spot on the secondary component. Actually, this version is limited in the following way: If the *simplex algorithm* is chosen for optimization and there is exactly one spot specified on star 2 then $(\varphi,\lambda,\rho,t_f)$ are always adjusted, i.e. there are four additional free parameters in the fit; no spot adjustment is permitted on star 1. Thus, there are two disadvantages: on the one hand, there are always four additional free parameters which usually is not a good idea at all. On the other hand it seems possible that the primary component cause some irregularities in the light curve and therefore it is star 1 which require spots on its surface.

4.2 Towards a complete control of spot fitting

For the reasons given above we concentrated on the definition of a flexible scenario for spot fitting which allows to model an arbitrary number of $n_s = n_{s1} + n_{s2}$ spots on both stars as in WD2 but there is no restriction on the choice of adjustable spot parameters. To fit m_s spots it is necessary to define m_s input vectors v_i $(1 \le i \le m_s)$ where

$$v_i := (\, j,\, k,\, s_1,\, s_2,\, s_3,\, s_4,\, k_1,\, k_2,\, k_3,\, k_4,\, \Delta s_1,\, \Delta s_2,\, \Delta s_3,\, \Delta s_4 \,)_i \,,$$

j,k indicate which spot k on star j, $(j=1,2)$, is to be adjusted, s_1 to s_4 are the corresponding spot parameters $(\varphi,\lambda,\rho,t_f)$ star, k_1 to k_4 specify whether a spot parameter is fixed, and Δs_1 to Δs_4 give the increments needed to construct the initial simplex (see Kallrath & Linnell 1987, p.355 for mathematical details). The definition of v_i supports the modelling of a wide range of scenarios, for instance to adjust the temperature factor t_f of the hotter star's spot, the longitude λ of the first, the latitude φ of the second, and the size ρ and temperature factor t_f of the third spot on the cooler star while keeping all other spot parameters as fixed ones.

In practice, far fewer spots than are given in the example should be used. Although it is an advantage to have as much flexibility as possible in the model it is not recommended to increase the number of spot parameters without caution. Even when modeling the eclipsing binaries *NH19* and *NH30* in the globular cluster NGC 5466 and allowing only for one dark spot $(t_f<1)$ on the cooler star to be adjusted, i.e. four additional free parameters, it turned out that it may change a light curve solution, i.e. the system parameters $(i,q,T_2,L_2/L_1,\ldots)$ drastically. Thus, spot fitting certainly requires a very careful discussion of questions of determinacy.

5 Conclusions

It is recommended to use both the *simplex algorithm* and *differential corrections* to derive light curve parameters. The former should be used for the initial search, and to seek for solutions in the whole parameter space. Furthermore, it is a helpful tool to support discussions of uniqueness and to perform parameter studies, e.g. for a grid of fixed mass ratios. If a parameter set close to a solution is found, switching to *differential corrections* can accelerate the rate of convergence and in addition yields the formal or statistical errors of the parameters which however should be interpreted with great caution. Spot fitting is now supported and solved with the *simplex algorithm*, and might be used in binary systems with a clear O'Connell effect during the later phase of light curve analysis. As is always the case with powerful analysis tools, the spot fitting option should be used with great caution.

Acknowledgements

The author gratefully acknowledges encouraging discussions with E.F.Milone during a two month research visit at the Physics & Astronomy Department of the University of Calgary in 1990, and thanks B.-C. Kämper and R.A. Breinhorst (Observatory, Bonn University) for helpful comments on this paper. The presentation of this paper at the modeling improvements meeting held under the auspices of Commission 42 during the XXIst General Assembly of the IAU was financially supported in parts by an IAU Young Astronomer Grant. Some initial work for this paper was done in autumn 1990 with the assistance from GKSS/FRG through the University of Bonn and the University of Calgary, and a grant from Natural Sciences & Engineering Research Council of Canada to E.F. Milone.

References

Andersen,J. & Gronbech,B.: 1975,'The close O-type eclipsing binary TU Muscae', *Astron. Astrophys.* **45**, 107-115.

Barone,F., Maceroni,C., Milano,C. & Russo,G. : 1988, 'The Optimization of the Wilson-Devinney method: An application to CW Cas', *Astron.Astrophys.* **197**, 347-353.

Barone,F., Milano,L. & Russo,G. : 1990, 'An Optimization Method for solutions of Close Eclipsing Binaries', in: Ibanoglu,I., *Active Close Binaries*, 161-188.

Breinhorst,R.A., Kallrath,J. & Kämper,B.-C. : 1989, 'Light Curve analysis and parameter study of the EB-type eclipsing binary V 836 Cygni', *Mon. Not. R. astr. Soc.* **241**, 559-573.

Budding,E. & Zeilik,M. : 1987, 'An Analysis of the Light Curves of Short-Period RS CVn stars: Starspots and Fundamental Properties', *Astrophys. J.* **319**, 827-835.

Davidge,T.J. & Milone,E.F. : 1984, 'A Study of the O'Connell effect in the Light Curves of eclipsing binaries', *Astrophys. J. Suppl.* **55**, 571-584.

Drechsel,H., Lorenz,R. & Meyer,P. : 1989, 'Solution of light curves with third light con tribution: the eclipsing binaries LY Aurigae and AH Cephei re-considered, *Astron.Astrophys.* **221**, 49-61.

Etzel,P.B. : 1981, 'A simple synthesis method for solving the elements of well-detached eclipsing systems' in: *Photometric and Spectroscopic Binary Systems*, ed. Carling,E.B. & Kopal,Z., 111-120.

Etzel,P.B. & Leung,K.-C. : 1990, 'Synthesis methods for eclipsing-binary light curves', in: Ibanoglu,I., *Active Close Binaries*, 873-879.

Fletcher,R. : 1970, 'A new approach to variable metric algorithms', *The Computer Journal* **13**, 317-322.

Gill,P.E., Murray,W. & Wright,M.H. : 1989, *Practical Optimization*, Academic Press, London, 8th printing.

Hill,G. : 1979, 'Description of an Eclipsing Binary Light Curve Computer Code with Application to Y Sex and the WUMA Code of Rucinski', *Publ. Dom. Astrophys. Obs.* **15**, 297-325.

Hill,G., Fisher,W.A. & Holmgren,D. : 1990, 'Studies of late-type binaries. IV. The physical parameters of ER Vul', *Astron.Astrophys.* **238**, 145-159.

Kallrath,J. & Linnell, A.P. : 1987, 'A new method to optimize parameters in solutions of eclipsing binary light curves', *Astrophys. J.* **313**, 346-357.

Kallrath,J. : 1987, *Analyse von Lichtkurven enger Doppelsternsysteme*, Diploma thesis, Bonn University Observatory.

Kallrath,J. & Kämper,B.-C. : 1992, 'Another look at the early-type eclipsing binary BF Aurigae', *Astron.Astrophys.*, submitted.

Kallrath,J., Milone,E.F. & Stagg,C.R. : 1992, 'Modeling of the Eclipsing Binaries in the Globular Cluster NGC 5466', *Astrophys. J.*, in press.

Kang,Y.W. & Wilson,R.E. : 1989, 'Least-Squares adjustment of spot parameters for three RS CVn binaries', *Astron.J.* **97**, 848-865.

Khaliullina,A.I. & Khaliullin,Kh.F. : 1984, 'Iterative method of differential corrections for the analysis of light curves of eclipsing binaries', *Sov. Astron.* **28**(2), 228-234.

Linnell,A.P. : 1984, 'A light synthesis program for binary stars I.', *Astrophys.J.Suppl.* **54**, 17-31.

Linnell,A.P. & Kallrath,J. : 1987, 'MR Cygni Revisited', *Astrophys. J.* **316**, 754-762.

Linnell,A.P., Peters,G.J. & Polidan,R.S. : 1988, 'An improved photometric analysis of SX Aurigae', *Astrophys. J.* **327**, 265-272.

Linnell,A.P. : 1992, 'Light Synthesis Modeling of Close Binary Stars', in: Milone,E.F. (ed.), *Modeling of Eclipsing Binary Stars*, Springer, New York, in press.

Marquardt,D.W. : 1963, 'An Algorithm for Least-Squares Estimation of Nonlinear Parameters', *SIAM J. Appl Math.* **11**, 431-441.

MicroMath : 1991, *Minsq - Nonlinear Curve Fitting and Model Development*, MicroMath Scientific Software, Manual.

Milone,E.F., Wilson,R.E. & Hrivnak,B.J. : 1987, 'RW Comae Berenices. III. Light Curve Solution and Absolute Parametere', *Astrophys. J.* **319**, 325-333.

Milone,E.F., Groisman,G., Fry,D.J.I.F. & Bradstreet.H. : 1991, 'Analysis and Solution of the Light and Radial Velocity Curves of the Contact Binary TY Bootis', *Astrophys. J.* **370**, 677-692.

Milone,E.F., Stagg,C.R. & Kurucz,R. : 1992, 'The Eclipsing Binary AI Phoenicis: New Results Based on an Improved Light Curve Analysis Program', *Astrophys. J.*, submitted.

Nelson,B. & Davis,W.D. : 1972, 'Eclipsing-binary solutions by sequential optimization of the parameters', *Astrophys. J.* **174**, 617-628.

Napier,W.M. : 1981, 'A simple approach to the analysis of eclipsing binary light curves', *Mon. Not. R. astr. Soc.* **194**, 149-159.

Nelder,J.A. & Mead,R. : 1965, 'A simplex method for function minimization', *The Computer Journal* **7**, 308-313.

Poe,C.H. & Eaton,J.A. : 1985, 'Star spot areas and temperatures in nine binary systems with late-type components', Astrophys. J. **289**, 644-659.

Plewa,T. : 1988, 'Parameter Fitting Problem for Contact Binaries', *Acta Astronomica* **38**, 415-430.

Press,W.H., Flannery,B.P., Teukolsky,S.A. & Vetterling,W.T. : 1989, *Numerical Recipes - The Art of Scientific Computing*, Cambridge University Press.

Price,W.L. : 1976, 'A controlled random search procedure for global optimization', *The Computer Journal* **20**(4), 367-370.

Schnell,A., Kallrath,J., Purgathofer,A. & Breinhorst,R.A. : 1992, 'Photoelectric BV Photometry of the W Ursae Majoris-Binary V 781 Tauri', *Astron. Astrophys. Suppl.*, in preparation.

Smak,J. : 1988, in: Swings,J.-P. (ed.), *REPORTS on ASTRONOMY - Transactions IAU*, Vol. **XXA**, 569-594.

Stagg,C.R., Milone,E.F., Sugars,B., McVean,J.R., Schiller,J. & Kallrath,J. : 1991, 'Light Curve and Radial Velocity Curve Analyses of the Eclipsing Binary H235 in the Open Cluster NGC 752', B.A.A.S. , in press, (abstract).

Walker,R.L. & Chambliss, : 1990, 'Analysis of the eclipsing binary ST Carinae using the simplex algorithm', *Astron.J.* **100**, 233-238.

Wilson, R.E. : 1979, 'Eccentric Orbit Generalization and Simultaneous Solution of Binary Star Light and Velocity Curves', Astrophys. J. **234**, 1054-1066.

Wilson,R.E. & Devinney,E.J. : 1971, 'Realization of Accurate Close-Binary Light Curves: Application to MR Cygni', Astrophys. J. **166**, 605-619.

Wilson,R.E. & Biermann,P. : 1976, 'TX Cancri - Which Component is Hotter ?', *Astron. Astrophys.* **48**, 349-357.

Wilson,R.E. & Rafert,J.B. : 1981, 'A note on the statistics of early-type close binaries', *Astrophys.Space Sci.* **76**, 23-33.

Wood, D.B. : 1971, 'An Analytic Model of Eclipsing Binary Star Systems', *Astron. J.* **76**, 701-710.

Yamasaki,A. : 1982, 'A spot model for VW Cephei', *Astrophys.Space Sci.* **85**, 43-48.

Zeilik,M., deBlasi,C., Rhodes,M. & Budding,E. : 1988, 'A Half-Century of starspot activity on SV Camelopardalis', *Astrophys. J.* **332**, 293-298.

The New Wilson Reflection Treatment and the Nature of BF Aurigae

W. Van Hamme
Department of Physics, University Park,
Florida International University, Miami, FL 33199, USA

Abstract

Light and radial velocity curves of the early-type close binary BF Aur are reexamined using a version of the Wilson-Devinney program which includes a new and accurate treatment of the reflection effect. This effect is expected to be important in BF Aur since it has two nearly equal components very close to contact. We compare solutions using the old simple and the new rigorous reflection treatment. Contrary to previously published solutions, we find BF Aur to be fully detached, but with radii between 95 to 97 per cent of their Roche lobe radii. A procedure for estimating monochromatic as well as bolometric limb-darkening coefficients for a linear and a logarithmic limb-darkening law is also outlined.

1. Introduction

Now entering its third decade of use, the Wilson-Devinney program (hereinafter called WD) is one of the most successful light and radial curve solution programs available to astronomers today. Although many reasons can be cited for this success, the main ones are probably the generality of the program, which enables it to handle many different types of binaries, and the solidity of the underlying physical model. Figures and surface gravity fields of the binary system components are accurately specified (stellar surfaces are equipotential surfaces, and gravity fields are obtained by differentiating the potential), while light and radial velocity curves are computed taking into account horizon and eclipse effects, reflection, gravity and limb darkening, and, if desired, star spots. Another hallmark of the program since its inception (Wilson and Devinney, 1971) is the simultaneous treatment of light curves in several passbands, which yields a single estimate for each adjusted wavelength-independent parameter. Wilson (1979) extended this idea further to allow the simultaneous solution of radial velocity and light curves, and added other features such as eccentric orbits and nonsynchronous rotation which increased the program's generality. This version, with an added star spot capability, is currently used worldwide. The program is again in the process of being revised. Some of the latest changes are cosmetic, ranging from streamlining the input data to optimizing parts of the program code. Other changes however are fundamental and represent improvements of the underlying physical model. While the development and implementation of these additions are solely the work of Wilson, a variety of tests, from code debugging to real binary applications, are being carried out by Wilson's collaborators, this author among them. In this talk we present the results of a new and improved reflection effect treatment (Wilson, 1990) and of the use of a two parameter limb-darkening law, using the close binary system BF Aur as an example.

2. The New Reflection Effect

A new rigorous and complete treatment of the mutual irradiation effects of two stars in a binary system was developed by Wilson (1990), and has been added to the WD program. The old version of the WD program treats the irradiating star as a point source from which light propagates according to an inverse square law and applies corrections due to penumbral and ellipsoidal effects to the flux received by the irradiated star. In the new program, the reflection effect is accurately calculated by using a model whose geometry and surface gravity fields are based on equipotential surfaces, allowing for all the usual binary star effects. Bolometric heating is separated from the wavelength-specific reradiation, and the interaction with other effects (gravity darkening, spots, etc.) is properly taken into account. Finally, the new reflection treatment allows for multiple reflection (star 1 heating star 2, thereby raising the temperature of star 2, and as a consequence, star 2 is now heating star 1 more than expected). The same geometric effects can be found in Hill's (1979) reflection treatment, but the radiative part of his approach is less rigorous and does not include multiple reflection.

There are a number of types of close binaries for which the old reflection treatment (inverse square law plus penumbral corrections) is expected to be more than adequate, as outlined by Wilson (1990). However, the simple reflection treatment is not expected to be very good in over-contact or non-synchronously rotating systems. A more rigorous and accurate computation is in order if we want to model these systems correctly. In slightly detached and semi-detached binaries the simple reflection model may be only moderately deficient, but in those cases even small errors could be important with regard to lobe filling vs. non-lobe filling solutions. So the new reflection treatment is expected to improve the modeling of binaries whose two components are close to or over fill their limiting lobes. Examples are the B-type W Ursae Majoris binaries, which presumably are in broken contact, and the hot O and B-type near contact systems. Light curve solutions often find these systems to be very close to contact and fail to distinguish between a marginally over-contact, or a slightly detached or semi-detached configuration.

Adding to the flexibility of the WD program, the old numerical treatment of reflection can still be accessed when it is deemed adequate, thus saving a substantial amount of computing time.

3. Limb-darkening Coefficients

The currently distributed WD program uses a linear limb-darkening law for which coefficients are usually estimated from the grid of model stellar atmospheres of Carbon and Gingerich (1969), or directly by interpolation in tables of linear limb-darkening coefficients by Al-Naimiy (1978). Coefficients obtained in this manner are, of course, monochromatic. In this section we outline our procedure for estimating *bolometric* limb-darkening coefficients as required by the new reflection treatment.

Using a customary notation, $I_\lambda(\mu)$ will be the monochromatic specific intensity of the emergent radiation in a direction $\Theta = \cos^{-1}\mu$ with the line of sight. We can write

$$I_\lambda(\mu)=I_\lambda(1)D_\lambda(\mu) \quad , \tag{1}$$

where

$$D_\lambda(\mu)=\frac{I_\lambda(\mu)}{I_\lambda(1)} \tag{2}$$

can be obtained from stellar atmosphere calculations, given the effective temperature, surface gravity, and chemical composition of the star. In many practical applications, such as programs which compute eclipsing binary light curves, the quantity $D_\lambda(\mu)$ is not calculated using a detailed stellar atmosphere program. Rather, it is approximated by some type of limb-darkening law. If a linear law is used, we have

$$D_\lambda(\mu)=1-x_\lambda(1-\mu) \quad , \tag{3}$$

in which the coefficient x_λ can be found by solving the equation

$$x_\lambda=3-6\left[\int_0^1 D_\lambda(\mu)\mu d\mu\right] \tag{4}$$

In this work limb-darkening coefficients were obtained using equation (4) and the tabulated D_λ values of Carbon and Gingerich (1969). The integral in (4) was numerically integrated using Gaussian quadratures.

The monochromatic emergent flux is given by

$$F_\lambda=2I_\lambda(1)\int_0^1 D_\lambda(\mu)\mu d\mu$$

which allows us to rewrite (1) as

$$I_\lambda(\mu)=\frac{F_\lambda}{2\int_0^1 D_\lambda(\mu)\mu d\mu}D_\lambda(\mu) \qquad (5)$$

Using fluxes from the atmosphere models by Kurucz, Peytremann, and Avrett (1974), equation (5) can be integrated over all wavelengths, yielding the bolometric specific intensity I(μ). Finally, the bolometric limb-darkening coefficient x can be obtained from the bolometric equivalent of equations (2) and (4).

Experiments were also done with a two-parameter limb-darkening law of the form

$$D_\lambda(\mu)=1-x_\lambda(1-\mu)-y_\lambda\mu\ln\mu \quad , \qquad (6)$$

as suggested by Klinglesmith and Sobieski (1970). In this case, the two limb-darkening coefficients x_λ and y_λ are given by

$$x_\lambda=18\left[\int_0^1 D_\lambda(\mu)\mu d\mu\right]-8\left[\int_0^1 D_\lambda(\mu)d\mu\right]-1 \qquad (7)$$

and

$$y_\lambda=36\left[\int_0^1 D_\lambda(\mu)\mu d\mu\right]-12\left[\int_0^1 D_\lambda(\mu)d\mu\right]-6 \quad , \qquad (8)$$

and the bolometric limb-darkening coefficients x and y can be obtained using the bolometric analogues of equations (7) and (8).

4. BF Aurigae Solutions

BF Aur is a 1.58 day period, double lined, close binary system with two nearly equal B5V type components. Mannino, Bartolini, and Biolchini (1964) (hereinafter called MBB) obtained light curves in U, B, and V, and Mammano, Margoni, and Stagni (1974) (hereinafter MMS) made spectroscopic observations and derived radial velocity curves. A solution of these radial velocity curves, combined with results of a

Russell and Merrill (1952) type photometric analysis, led MMS to the conclusion that the two components are in contact with their Roche lobes. Schneider, Darland, and Leung (1979) (hereinafter referred to as SDL) used an earlier version of the Wilson-Devinney (1971) program to solve MBB's light curves. They presented two alternate semi-detached solutions, pointing out the uncertainty of the mass ratio q. A solution with q = 0.831 and a lobe filling primary component appears to have nearly the same sum of squared residuals as a solution with q = 1.20 and a lobe filling secondary. They further claimed that the semi-detached solutions were significantly better than a detached solution with q set to the spectroscopic value 1.00 obtained by MMS.

The reflection effect in BF Aur was expected to be important because of the proximity of the surfaces of its components and their nearly equal temperatures. Correctly modeling this effect might then help to determine more clearly the true configuration of the system, i.e. whether it is marginally over-contact or near-contact but detached or semi-detached. So we decided to reanalyze the V and B light curves of MBB and the radial velocity curves of MMS with the old and new reflection treatments.

Differential magnitudes were converted to light values and normal point light curves were made, typically using approximately two to three observations per normal. Individual normal point weights were assigned equal to the number of observations per normal. Phases were computed with the ephemeris

Min. (JD Hel.) = 2440628.3644 + 1.5832208 E

obtained by Guarnieri, Bonifazi, and Battistini (1975). Radial velocity phases were computed with the same ephemeris. Curve-dependent weighting for the light curves was done in the following manner: four short phase intervals (average lengths $\sim 0.^{P}09$) around maximum light were selected and linear least squares fits to the light values in these intervals were made. Geometric means of the standard deviations of these fits were 0.0094 and 0.0073 light units in V and B respectively, and these were used as curve-dependent weights. Comparison of these standard deviations with the standard deviations of linear least squares fits to the light values at the bottom of the primary minimum (0.0086 in V and 0.0074 in B) indicated that the scatter in both light curves was proportional to the square root of light level, hence the level-dependent weight was chosen proportional to ℓ^{-1}. We adopted a temperature T1 = 15600 K for the primary component in accordance with the B5V spectral type. The gravity-darkening exponents and bolometric albedos selected were appropriate for stars with radiative outer envelopes. Our procedure to select limb-darkening coefficients was outlined in Section 3.

Individual radial velocities and weights were used as listed in MMS. No normal points were made. In order to apply curve-dependent weighting, we proceeded as follows: first, we used the WD differential corrections program to solve the two velocity curves with equal weighting and the point source condition (by setting the potentials to very large values so that proximity effects become negligible); then we kept the mass ratio from this solution fixed and separately solved the primary and secondary radial velocity curves, adjusting only the systemic velocity and semi-major axis. The resulting large standard deviations were 52.8 and 55.2 km s^{-1} for the primary and secondary velocity curves respectively, and these were used as the basis for computing curve-dependent weights in the WD program. They reflect the rather poor quality of the radial velocity curves (quality rating d in Batten, Fletcher, and MacCarthy, 1989). (Note: the star labeled 1 by MMS is eclipsed at secondary eclipse, so we reversed the components, and component 2 in MMS became component 1 in this study.)

Pure photometric, pure spectroscopic, and simultaneous photometric-spectroscopic solutions were made, but only the results of the simultaneous solutions are presented in Table 1. Figure 1 compares the observed and computed light curves according to the simultaneous solution with the new reflection treatment and for double reflection, and Figure 2 shows the corresponding radial velocity curves.

A simultaneous light and radial velocity curve solution using a two-parameter limb-darkening law of the form (6) was made. The results are also listed in Table 1 and are compared to our other solutions in Section 5.

5. Discussion

All solutions listed in Table 1 correspond to detached configurations with the two components within, but close to, their limiting Roche lobes. This is not in agreement with the SDL results, as they found a semi-detached configuration to be significantly better than a detached one. Light curve solutions were made in program mode 4 (with a primary component constrained to fill its Roche lobe) and mode 5 (Roche lobe filling constraint for the secondary), using initial parameters according to the SDL solutions. In both cases, the differential corrections program quickly converged to a solution with a mass ratio q closer to unity (about 0.95 in mode 4, 1.05 in mode 5). Subsequently, after relaxing the contact constraint and continuing the differential corrections program in mode 2, we found a detached solution with a 2% reduced sum of squared residuals. Because the mass ratio q is close to unity, and photometrically not well determined, we decided to make simultaneous photometric-spectroscopic solutions to find a q value consistent with both kinds of observations. (To avoid over burdening the reader with solutions, only the simultaneous results are listed in Table 1.)

Figure 3 shows the fraction of total light due to reflection [the total light with reflection (albedos A1 = A2 = 1) minus the total light without reflection (A1 = A2 = 0), normalized to the maximum light without reflection] versus phase. The dotted line was obtained using the simple reflection treatment, while the dashed and solid lines represent the results from the new detailed reflection treatment, with single and double reflection respectively. Although BF Aur has two very similar components nearly filling their Roche lobes, reflection never amounts to more than 1.5% of total light. Consequently, the details of the reflection computation have little effect on the outcome of the solutions. Some of the individual parameters, such as the temperature of the secondary component, appear to depend marginally on the choice of reflection treatment and on the number of reflection iterations, but the resulting absolute dimensions are hardly affected. The use of the old reflection treatment can continue without seriously affecting the outcome of light curve solutions, at least for systems similar to BF Aur. However, our experience using the new reflection program is, so far, limited to this one system, and future work on many other binaries will be required in order to draw any firmer conclusions.

In all of our solutions, the secondary component is slightly more massive than the primary. The secondary is also slightly larger than the primary, except when a two-parameter limb-darkening law is used. In this case, the less massive primary is slightly larger than the secondary. The difference, however, is small. Regardless of the details of the solution, BF Aur appears to be a detached system with two very similar components filling approximately 95 to 97 per cent of their Roche lobes. New spectroscopic observations and better radial velocities are needed to confirm the value of the mass ratio. The probable errors of this parameter listed in Table 1 are, of course, internal errors of the solution, and may under estimate its real uncertainity, given the poor quality of the radial velocity curves.

A project to construct tables of monochromatic and bolo-metric limb-darkening coefficients [for limb-darkening laws of the form (3) and (6)] is currently being undertaken by the author of this paper. Fluxes and specific intensities will be obtained using the most recent version of the ATLAS stellar atmosphere program of Kurucz (1979) available.

Acknowledgement

Travel support to attend the International Astronomical Union General Assembly was provided by an award from the National Aeronautics and Space Administration administered by the American Astronomical Society. Additional funding was provided by the College of Arts & Sciences and the Division of Sponsored Research of Florida International University. This work was also supported by a NASA Interinstitutional Space Research Grant, administered by the Florida Space Grant Consortium.

References

Al-Naimiy, H.M., 1978, Ap&SS 53, 181
Batten, A.H., Fletcher, J.M., and MacCarthy, D.G., 1989, *Eight Catalogue of the Orbital Elements of Spectroscopic Binary Systems*, Publ. Dom. Astrophys. Obs. XVII
Carbon, D., and Gingerich, O., 1969, *Theory and Observation of Normal Stellar Atmospheres*, ed. O. Gingerich (MIT, Cambridge, MA), p. 377
Guarnieri, A., Bonifazi, A., and Battistini, P., 1975, A&AS 20, 199
Hill, G., 1979, Publ. Dom. Astrophys. Obs. 15, 297
Klinglesmith, D.A., and Sobieski, S., 1970, AJ 75, 175
Kurucz, R.L., 1979, ApJS 40, 1
Kurucz, R. L., Peytremann, E., and Avrett, E.H., 1974, *Blanketed Model Atmospheres for Early-Type Stars*, Smithsonian Institution Press, Washington, D.C.
Mammano, A.M., Margoni, R., and Stagni, R., 1974, A&A 35, 143 (MMS)
Mannino, G., Bartolino, C., and Biolchini, R., 1964, Mem. Soc. Astron. Ital. 35, 371 (MBB)
Russell, H.N., and Merrill, J.E., 1952, Contr. Princeton U. Obs. No. 26
Schneider, D.P., Darland, J.J., and Leung, K.C., 1979, AJ 84, 236 (SDL)
Wilson, R.E., 1979, ApJ 234, 1054
Wilson, R.E., 1990, ApJ 356, 613
Wilson, R.E., and Devinney, E.J., 1971, ApJ 166, 605

TABLE 1

Simultaneous Light and Radial Velocity Curve Solutions for BF Aur

	MREF = 1[a] NREF = 1[b]	MREF = 1[a] NREF = 1[b]	MREF = 2[a] NREF = 1[b]	MREF = 2[a] NREF = 2[b]
$a(R_\odot)$	12.32 ± 0.21 p.e.	12.33 ± 0.21 p.e.	12.29 ± 0.21 p.e.	12.30 ± 0.21 p.e.
F1	1.0	1.0	1.0	1.0
F2	1.0	1.0	1.0	1.0
ϕ	$0.^P0005 \pm 0.^P0001$	$0.^P0005 \pm 0.^P0001$	$0.^P0005 \pm 0.^P0001$	$0.^P0005 \pm 0.^P0001$
$V_\gamma(kmS^{-1})$	+8.0 ± 2.8	+8.0 ± 2.8	+8.0 ± 2.8	+8.1 ± 2.8
i	$84.^\circ41 \pm 0.^\circ11$	$84.^0.244 \pm 0.057$	$84.^\circ558 \pm 0.^\circ059$	$84^\circ.621 \pm 0.^\circ068$
g1	1.0	1.0	1.0	1.0
g2	1.0	1.0	1.0	1.0
T1	15600 K	15600 K	15600 K	15600 K
T2	15341 ± 55 K	15237 ± 46 K	15310 ± 44 K	15400 ± 38 K
A1	1.0	1.0	1.0	1.0
A2	1.0	1.0	1.0	1.0
$\Omega 1$	3.909 ± 0.027	3.852 ± 0.020	3.898 ± 0.025	3.955 ± 0.022
$\Omega 2$	3.844 ± 0.030	3.912 ± 0.025	3.875 ± 0.031	3.844 ± 0.027
q	1.0169 ± 0.0087	1.0177 ± 0.0067	1.0231 ± 0.0086	1.0274 ± 0.0099
$L1/(L1+L2)_V$	0.490	0.517	0.497	0.476
$L1/(L1+L2)_B$	0.491	0.518	0.498	0.477
X1(V)	0.30	0.44	0.30	0.30
Y1(V)	-	0.21	-	-
X1(B)	0.37	0.55	0.37	0.37
Y1(B)	-	0.27	-	-
X1(BOL)	0.60	0.695	0.60	0.60
Y1(BOL)	-	0.180	-	-
X2(V)	0.30	0.45	0.30	0.30
Y2(V)	-	0.21	-	-
X2(B)	0.38	0.56	0.38	0.38
Y2(B)	-	0.27	-	-
X2(BOL)	0.61	0.696	0.61	0.61
Y2(BOL)	- -	0.177	-	-
l3(V)	0.000	0.000	0.000	0.000
l3(B)	0.000	0.000	0.000	0.000
$\Sigma w(Res)^2$	0.01182	0.01182	0.01170	0.01172
σ	0.005224	0.005226	0.005198	0.005203

Table 1 Continued

	MREF = 1[a] NREF = 1[b]	MREF = 1[a] NREF = 1[b]	MREF = 2[a] NREF = 1[b]	MREF = 2[a] NREF = 2[b]
		Relative Radii (R/a)		
r_1(pole)	0.3395 ± 0.0029	0.3460 ± 0.0021	0.3413 ± 0.0027	0.3354 ± 0.0022
r_1(point)	0.4121 ± 0.0085	0.4334 ± 0.0077	0.4185 ± 0.0084	0.4023 ± 0.0060
r_1(side)	0.3542 ± 0.0035	0.3619 ± 0.0025	0.3564 ± 0.0032	0.3495 ± 0.0026
r_1(back)	0.3786 ± 0.0046	0.3890 ± 0.0033	0.3818 ± 0.0043	0.3727 ± 0.0035
$\underline{r}_1$[c]	0.3583	0.3667	0.3607	0.3533
%[d]	94.7%	96.9%	95.4%	93.6%
r_2(pole)	0.3496 ± 0.0042	0.3422 ± 0.0032	0.3477 ± 0.0042	0.3525 ± 0.0041
r_2(point)	0.4397 ± 0.0182	0.4150 ± 0.0097	0.4315 ± 0.0157	0.4492 ± 0.0213
r_2(side)	0.3660 ± 0.0051	0.3571 ± 0.0038	0.3637 ± 0.0051	0.3693 ± 0.0051
r_2(back)	0.3933 ± 0.0072	0.3814 ± 0.0051	0.3900 ± 0.0070	0.3975 ± 0.0072
$\underline{r}_2$[c]	0.3707	0.3610	0.3681	0.3742
%[d]	97.2%	94.7%	96.4%	97.9%
		Absolute Dimensions		
$M_1(M_\odot)$	4.97	4.98	4.92	4.92
$R_1(M_\odot)$	4.42	4.52	4.43	4.35
$\log(L1/L_\odot)$	3.01	3.03	3.01	2.99
$M_2(M_\odot)$	5.06	5.07	5.03	5.05
$R_2(R_\odot)$	4.57	4.45	4.52	4.60
$\log(L_2/L_\odot)$	3.01	2.98	3.00	3.02

[a] MREF = 1 Old Reflection Treatment
MREF = 2 New Reflection Treatment
[b] NREF Number of Reflection Iterations
[c] Radius of a sphere with equal volume
[d] [$\underline{r}$/$\underline{r}$(Roche lobe)]100

Fig. 1 B, V light curves of BF Aur by Mannino, Bartolino, and Biolchini, and our solution curves (for the simultaneous light and radial velocity solution with detailed and double reflection).

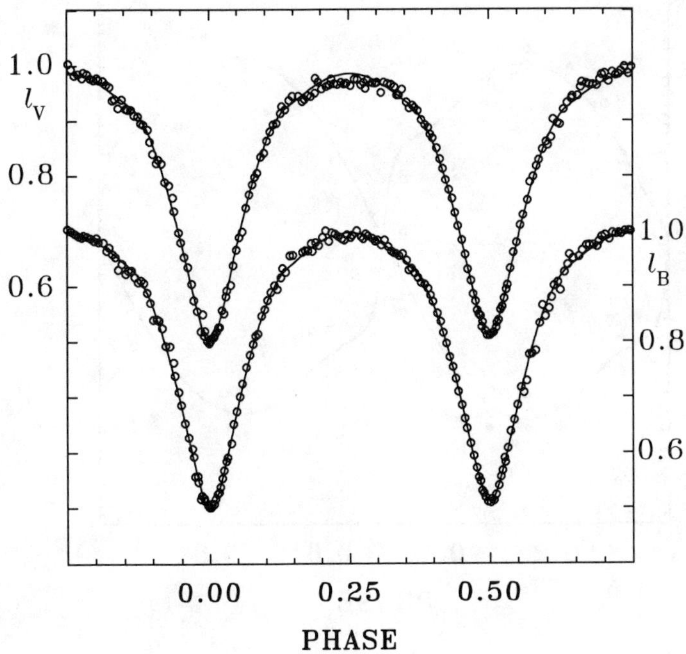

Fig. 2 Radial velocity curves of BF Aur by Mammano, Margoni, and Stagni, and our fitted curves for the simultaneous light and radial velocity curve solution (filled circles are for component 1, open circles for component 2).

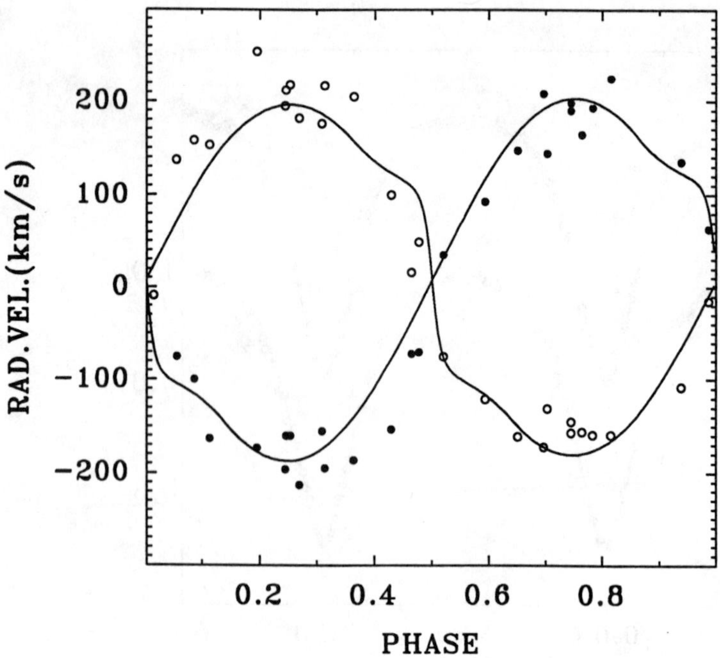

Fig. 3 Reflected total light in B (top) and V (bottom) for BF Aur. The dotted lines were obtained with the simple point source irradiation model, the dashed (single reflection) and solid lines (double reflection) with the detailed reflection model.

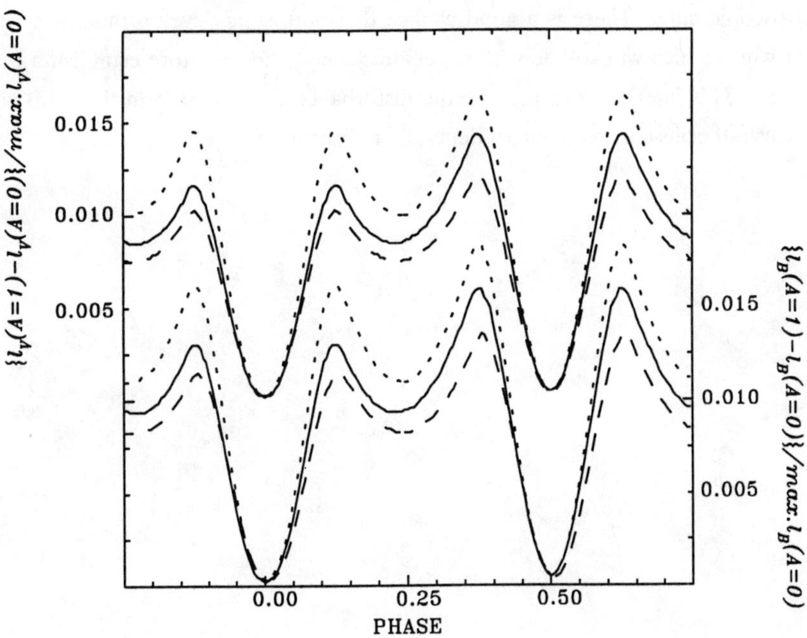

Discussion

Q. (J. Kallrath to W. Van Hamme): I would like to stress that BF Aurigae is a really interesting system and it deserves further attention and improved photometric and spectroscopic data. There is a good chance that both stars develop radiatively driven stellar winds which will collide with supersonic speed and therefore establish a contact interface. This interface might cause the disturbance at phase 0.25 in the U, B, and V light curves because of wavelength-dependant absorption.

Binary Star Rotation from Line Profiles

J. Mukherjee
Astronomy Department, University of Florida, Gainesville, FL 32611, USA
G.J. Peters
Space Sciences Center, University of Southern California, University Park, Los Angeles, CA 90089-1341, USA
R.E. Wilson
Astronomy Department, University of Florida, Gainesville, FL 32611, USA

ABSTRACT

Rotation measures for binaries and single stars need to be put on a firm basis. From an astrophysical point of view, asynchronous rotation in Algol type binaries is mainly a consequence of mass transfer, and thus can provide evolutionary information. Also, rotation is a neglected area of stellar astronomy, especially on the observational side, and it is appropriate to develop our knowledge simply because unexpected benefits nearly always accompany rapid observational advances into unexplored territory. This contribution includes results on RY Per, RW Mon, DM Per, RY Gem, TU Mon and Algol (Table 1). F_1 is the ratio of spin angular speed to mean orbital angular speed for the primary star while F_2 is the corresponding ratio for the secondary star. F_1= 1 signifies synchronous rotation. Table 1 lists the values of F_1 obtained from three methods. F_1(dc) refers to the value obtained by the differential correction method while F_1(simplex) refers to the value obtained from the simplex method. F_1(lc) is the photometric F_1 value. The published spectroscopic F_1 values in Table 1 are averages found by Van Hamme and Wilson (1990, A.J 100, 1981) from the listed references. The last column in Table 1 gives the equatorial velocity corresponding to the F_1(dc) value. The fitted spectral line profiles are shown in Figures 1, 2, and 3. A more complete list of binaries observed to date by G.J.Peters is given in Table 2. In this table the term "noisy lines" refers to those for which the spectral distributions have large scatter. Any blending is usually with lines of the same star because Algol-type secondary stars are much less luminous than the primaries. The term "clean lines" refers to those which are reasonably free of scatter and blending. The resolution of the CCD as it was used is 0.15 Angstroms per pixel.

Traditional methods for determining stellar rotation rates from photospheric spectral line profiles are subjective, do not provide standard error estimates, and neglect most or all other broadening mechanisms. Essentially impersonal results of greatly improved reliability are now potentially attainable, due to advances in the areas of observations, modeling, and parameter adjustment. Improvements in observations are coming along naturally because of developments in detectors, such as CCD's and reticons. It is important for modeling and parameter adjustments to keep pace with the improved observations. At the University of Florida, we have developed a model which produces computed line profiles (for theory, see Mihalas, D. 1978. Stellar Atmospheres. San Francisco, W.H. Freeman and Company) which are Doppler broadened by superposition of local intrinsic profiles over the surfaces of model stars (WD model). Effects included are thermal Doppler broadening, turbulence, damping, rotation, gravity darkening, limb darkening, reflection, tidal distortion, eclipses, and instrumental broadening. Phase smearing will be added soon. Among the parameters are the usual binary star parameters (including the F_1 and F_2 rotation parameters and the surface temperature (T_{eff})), turbulent velocity (V_{turb}), effective damping constant (Γ), and the number of absorbers in the line of sight (Nf). Our simplex and differential corrections programs for parameter fitting are now being used on CCD line profiles of additional Algol primaries recently observed by G.J.Peters.

Part of this research was supported by grant NGT 40015 from the Florida Space Grant Consortium.

Table 1 : F_1 values by three methods

STAR	F_1(dc)	F_1(simplex)	F_1(lc)	F_1(sp,publ)	V_e *sin i* (km/sec)
RY Gem	8.76 ± 0.29	8.76	14.42* ± 0.79	-	117
RY Per	9.62 ± 0.11	9.63	10.82** ± 0.30	9.11†	302
RW Mon	1.27 ± 0.03	1.27	4.99* ± 0.27	-	68
Algol	1.14 ± 0.04	1.13	-	1.04‡	57
DM Per	2.11 ± 0.03	2.08	-	-	152
TU Mon	4.69 ± 0.08	4.67	-	-	272

* Van Hamme, W., and Wilson, R.E. (1990). Astr. J. **100**, 1981

** Van Hamme, W., and Wilson, R.E. (1986). Astr. J. **92**, 1168

† Van den Heuvel, E.P.J. (1970). In *Stellar Rotation*, edited by A. Slettebak (Reidel, Dordrecht), p. 178

† Van Hamme, W., and Wilson, R.E. (1986). Astr. J. **92**, 1168

‡ Van den Heuvel, E.P.J. (1970). In *Stellar Rotation*, edited by A. Slettebak (Reidel, Dordrecht), p. 178

‡ Rucinski, S.M. (1979). Acta Astr. **29**, 339

‡ Olson, E.C. (1984). Publ. Astr. Soc. Pac. **96**, 376

‡ Tan, H. (1986). Chinese Astr. Astrophys. **10**, 54

Table 2 : Stars observed by G.J. Peters

CLEAN LINES		BLENDED & NOISY LINES	
RZ Cas	Z Ori	KX And	RZ Oph
TV Cas	DM Per	IM Aur	DN Ori
XX Cep	IZ Per	SX Aur	AQ Peg
U CrB	RY Per	RX Cas	AW Peg
VW Cyg	β Per	SX Cas	HU Per
WW Cyg	Y Psc	SW Cyg	RT Per
V367 Cyg	U Sge	SX Dra	RW Per
W Del	V505 Sgr	RX Gem	FV Sco
AI Dra	HU Tau	SV Gem	V393 Sco
CX Dra	RW Tau	TT Hya	RZ Sct
S Equ	λ Tau	TT Lyr	DL Vir
RY Gem	TX UMa	β Lyr	
DH Her	RS Vul	AU Mon	
TU Mon	Z Vul	UX Mon	

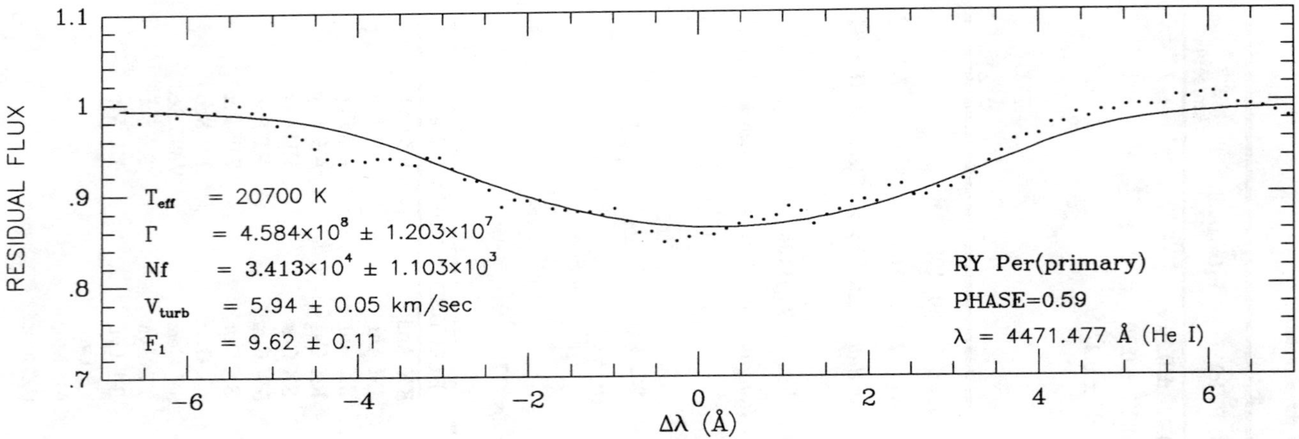

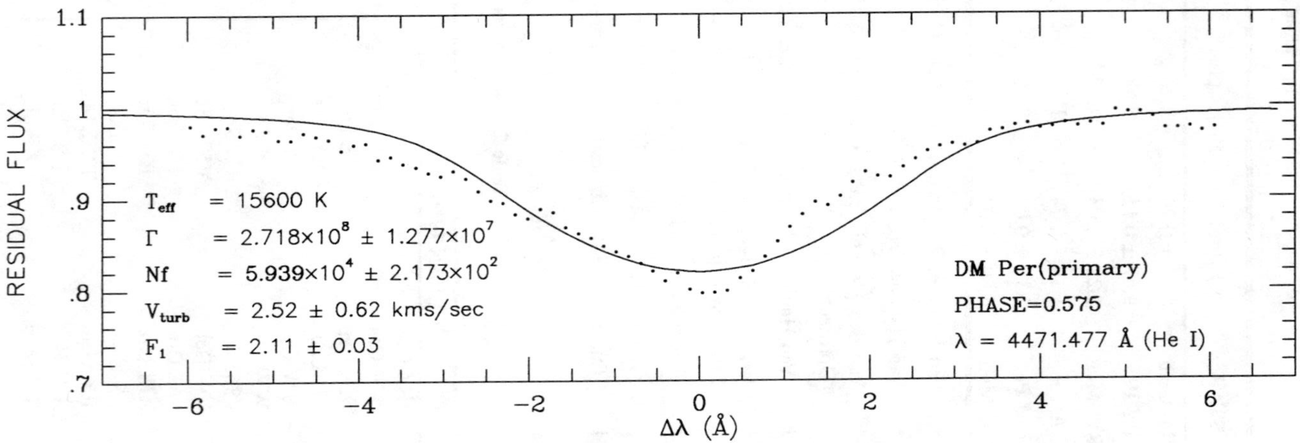

FIG. 1. Peter's line profiles (dots) for RY Per and DM Per and our fitted curves

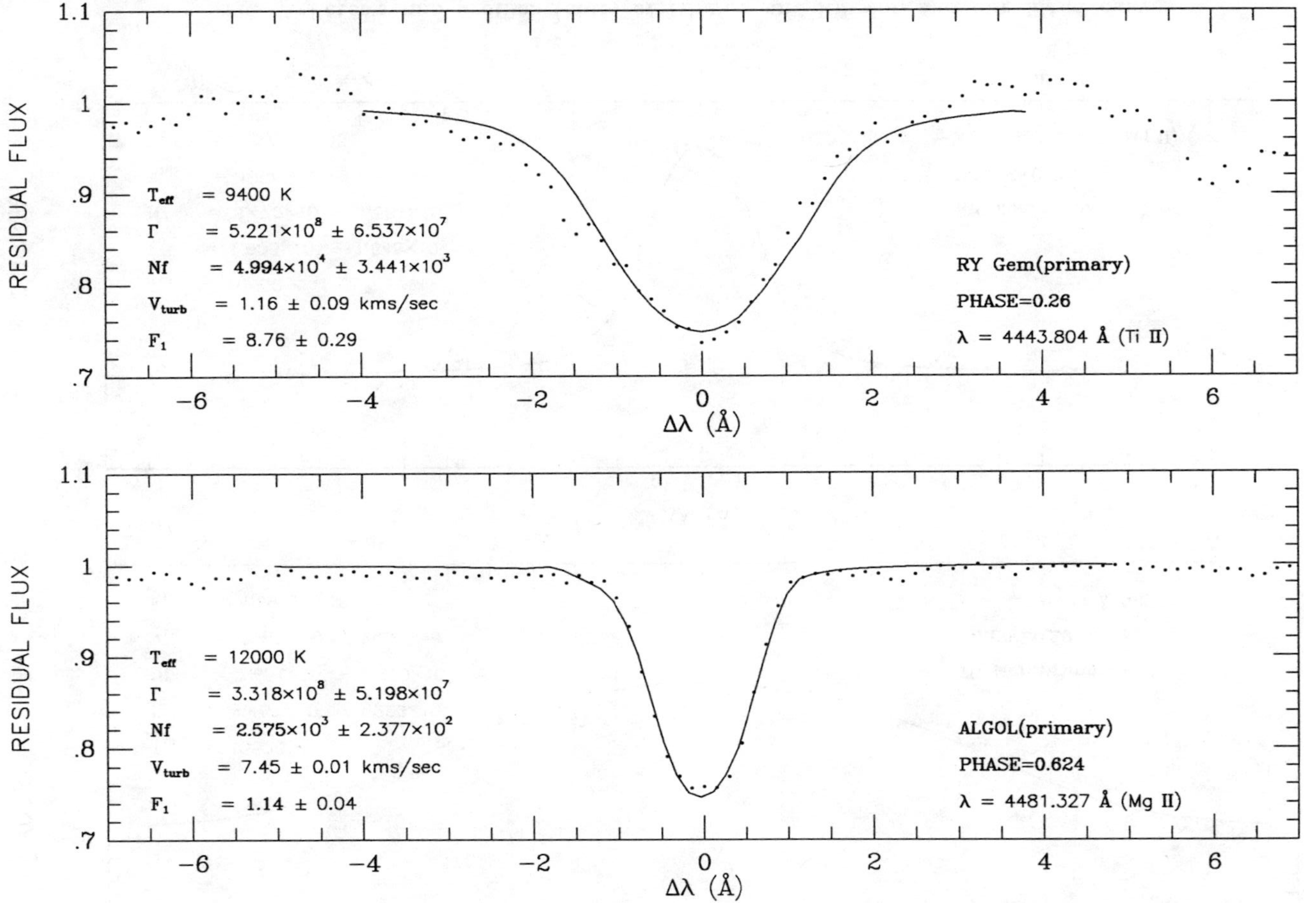

FIG. 2. Peter's line profiles (dots) for RY Gem and Algol and our fitted curves

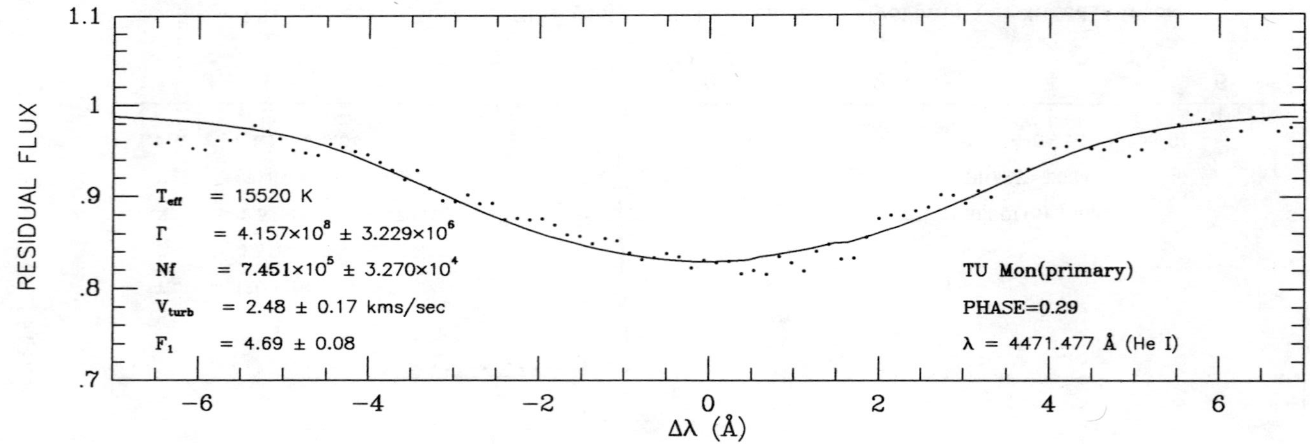

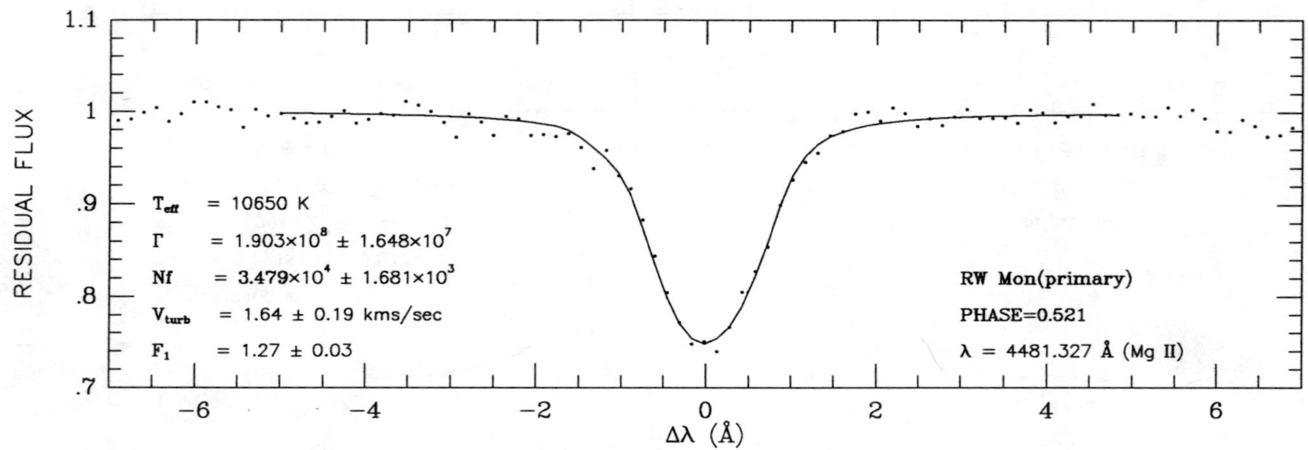

FIG. 3. Peter's line profiles (dots) of TU Mon and RW Mon and our fitted curves

Improvements to the Wilson-Devinney Code on Computer Platforms at the University of Calgary[1]

C.R. Stagg and E.F. Milone
Rothney Astrophysical Observatory, University of Calgary,
2500 University Drive NW, Calgary, Alberta, Canada T2N 1N4

Abstract

We describe improvements in light curve modeling over the past five years at the University of Calgary, and the current status of the effort. The work has concentrated on the atmospheres option in the Wilson-Devinney program, the uniqueness determination question, and code portability. We also describe a number of useful tools to facilitate adjustments and the examination of results and consider areas for further development.

Introduction

In the Beijing Colloquium, Milone (1989) suggested several ways in which the Wilson-Devinney eclipsing binary modeling program (hereafter referred to as *WD*) could be made more powerful and useful for eclipsing binary studies. A brief progress report on the improvements carried out at the University of Calgary was provided at the Pacific Rim meeting by Milone (1991). Here we discuss these improvements from a slightly different perspective and indicate avenues for further progress.

---

[1] **Publications of the Rothney Astrophysical Observatory No. 63**

The basic program on which the improvements have been implemented is that of Wilson and Devinney (1971) as modified to 1985 (Wilson and Sofia 1976; Wilson and Biermann 1976; Wilson 1983). The main improvements have been as follows:

* Optimization and limited vectorization for the Cyber 205 (Milone *et al.* 1989);

* Replacement of the *ATM* subroutine with a new version based on Kurucz atmospheres (Milone, Stagg, and Kurucz 1991);

* Inclusion of log **g**, specified in DC and LC input by an appropriate value of the **IFAT** parameter;

* Standard **UBV** and **uvby** passband responses, automatically selected by setting the wavelength equal to specific values;

* Empirical corrections for predicted **U**, **u**, and **b** fluxes may be applied;

* Extrapolation beyond the lowest temperatures in the Kurucz tables (5500 K);

* Decoupling of limb darkening parameters $\mathbf{x}_1$ and $\mathbf{x}_2$ for contact systems (Milone *et al.* 1991).

In addition to these code improvements, several procedural improvements were carried out: the first optimized spot determination was attempted by Milone, Hrivnak, and Wilson (1987) and carried further for 3 spot parameters by Groisman (1990) and in Milone *et al.* (1991). A series of tests of uniqueness and model robustness performed on the systems RW Com and TY Boo are also described in these references.

Parallel Processing

Parallel processing has turned out to be an extremely valuable tool for binary star modelling because a large number of light curves can be processed simultaneously; for a system like AI Phoenicis, for which we have analyzed UBVRI, uvby, four far UV partial light curves and two radial velocity curves, the running time on a Myrias computer is of the same order and indeed slightly less than the single processor Cyber 205. Thus the massive parallelism of the Myrias machine more than makes up for its slower processors. We are now able to do a dozen DC runs in only a few hours at most, something that was previously impossible using slower machines. To take advantage of the quick turnaround time, procedures have been developed which quickly interpret the DC output, check for problems, and apply corrections to the DC input file.

New Procedural Improvements

The Wilson-Devinney code was initially designed <u>not</u> to carry out automatic iteration in order to force the user to examine each and every output to avoid the generation of physically unrealistic parameters and conditions. Indeed, the use of subsets of uncorrelated parameters often results in a rapid multiplication of choices for successive runs, and sometimes the errors for the subset predictions are not greatly different, a condition which makes the choice of a succeeding adjustment problematic. We have therefore approached the iteration question warily.

After each run with the differential corrections program, we examine the resulting parameter adjustments in each of the subsets and decide on the necessary changes. Two programs, *RAD* and *READ2*, have been created by CRS to aid in this process. *RAD* corrects input and *READ2* interprets output: *RAD* takes the old *DC* input file, applies parameter corrections, and creates a new *DC* input file and also obtains new limb-darkening parameters from the tables of Wade and Rucinski, which we have on-line, if desired; *READ2* provides a brief summary of parameter corrections and correlation coefficients from the *DC* output and also summarizes corrections and standard errors for each parameter set, correlation coefficients, and stellar radii. After eliminating those parameter sets whose members are correlated, we select the parameter set whose corrections give the lowest predicted SSR, and have *RAD* apply them. Resubmission, from the receipt of *DC* output to the submission new *DC* input, typically takes 2-3 minutes.

To provide for quick visualization of results, the following procedures have been adopted for running *LC* and treating output: the program *MAKEFIL* creates *LC* input files from the *DC* input file. Separate files are created for each passband. A 1000 point input file typically takes about 2 minutes on the Myrias - through adaptation of *LC* for parallel processing; the output of *LC* is downloaded to a PC where the software package *TDPREVU* plots the observed and predicted light curves and from which the screen display can be copied to printer; output also can be downloaded to a main frame (Honeywell *MULTICS*) where the program *WDGRAPH* can be used to plot observed and predicted light curves, which can then be sent to the screen or to a postscript printer and retrieved at batch control at the Academic Computer Centre (in about 15 minutes).

Uniqueness Determination

In general, we have explored the uniqueness of a solution by perturbing the solution elements by ± 10% to see if another, deeper minimum lay within range. This work was undertaken systematically for TY Boo first by G. Groisman in an MSc. thesis (Milone *et al.* 1991). However, a broader approach is now possible.

A simplex-enfolding version of our revised Wilson-Devinney code has now been developed (Kallrath, Milone & Stagg 1992), which greatly aids uniqueness trials. It uses simplex algorithms to sample regions of parameter space while efficiently discovering the absolute minimum. It has a number of advantages over the differential corrections version alone:

* automatic adjustment of spot parameters;
* exploration of a greater volume of parameter space in a single run;
* *less* user interaction is required;
* calculation of Roche surfaces, which can then be plotted together with inner and outer critical surfaces.

However, DC has the following advantages:

* greater accuracy;
* faster convergence;
* *more* user interaction is required;
* suitability for parallel processing.

We have used simplex to explore parameter space by setting initial simplices at widely differing initial parameter configurations. We have in all cases checked the converged solution with runs on *DC*. It is not entirely surprising that some differences in the converged solution should be discovered because the convergence technique is different, and not all light and radial velocity curves present well-conditioned data sets. Small differences occasionally found are attributable to indeterminacies inherent in the data.

Since the simplex version is not parallelizable, this version of the code is being run on the IBM RS6000 system and on a Sun SPARCstation. We compare the run time of

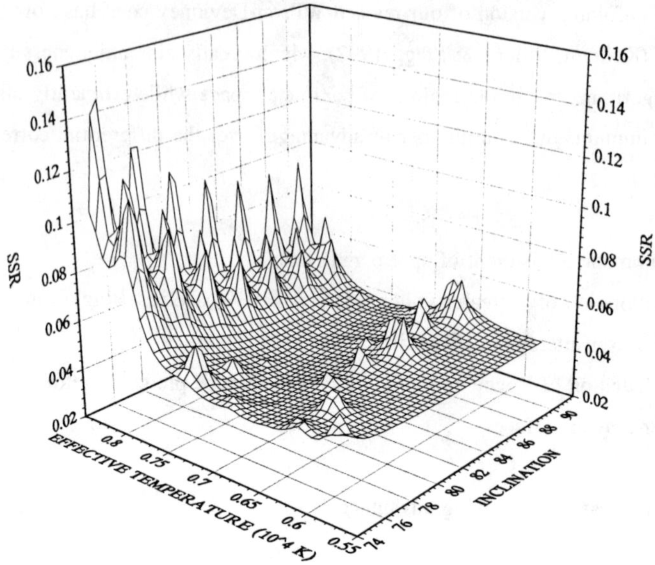

Fig. 1 - Σwr^2 *vs.* T and i in the modeling of NGC5466-NH30. The representation here and in Figs. 2-4 were made created with the commercial package *GRAPHTOOL.*

these platforms below. The automatic spot adjustment capability was first used in our investigation of the eclipsing binary stars in the globular cluster NGC 5466; previously, we undertook a spot optimization program through repeated trials with DC, altering each parameter in turn and determining the minimum error by parabolic fit of the Σwr^2 figure of merit. This quantity is also used to test the uniqueness of the robustness and uniqueness of all solutions. In Fig. 1, the quantity Σwr^2 is the third dimension in the slice of parameter space containing ι and $\mathbf{T_2}$ for the difficult contact system NH30 in the globular cluster NGC 5466. The data in this example are from CCD data from Mateo *et al.* (1990). The highland areas are somewhat misleadingly presented as mountain peaks because of the surface contour mapping scheme provided with this package, but the data cluster around the converged minimum. This graph is the result of 11 simplex runs at 30 iterations per simplex, with starting points covering the ranges: $6000 \leq T_2 \leq 8000$ K and $75° \leq i \leq 90°$. The deep valley at constant $\mathbf{T_2}$ indicates that the inclination is not well determined in this subset of parameter space. The converged *DC* solution elements and their *rms* uncertainties from the final full-set solution are shown superimposed on the contours obtained in successive trials with the simplex version in Fig. 2. Note that the *WD* and *SWD* results are not significantly different. The shallowness of the contours and the large *DC* error bars result from the sparse and noisy data thus far available for this faint (>19th magnitude) system. A fuller discussion of the modeling done on this system can be found in Kallrath, Milone, and Stagg (1991), which represents the first use of the simplex-enfolded version of our updated *WD* program. Whereas the contact parameter f = 0.94 for the NH30 model, the second example shows better conditioned data for a contact system with a much smaller contact parameter (0.4). The system is H235 in the intermediate age open cluster NGC 752. Here despite two nearly equally likely solutions, the parameters are quite well-determined because both solutions call for only very small temperature difference between components.

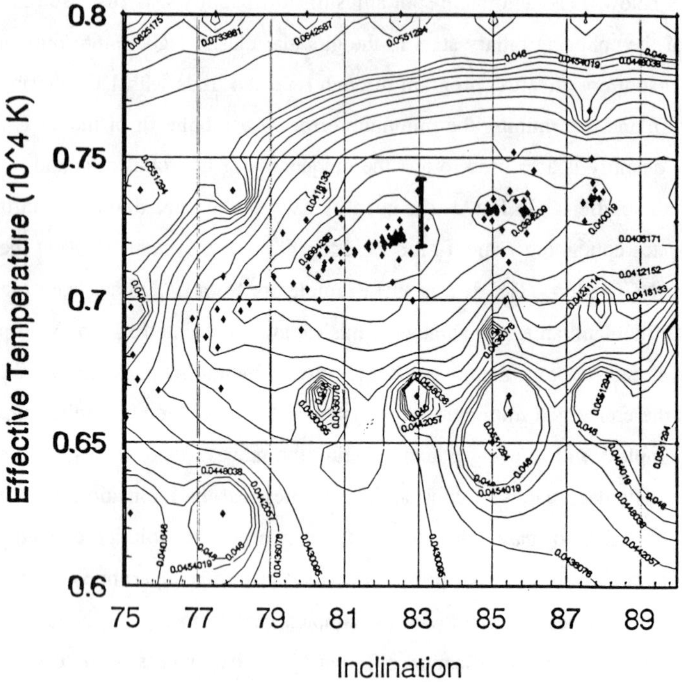

Fig. 2 - Contour plot of Σwr^2 *vs.* T and i in the modeling of NGC5466-NH30, showing agreement, within wide errors, of *WD* and simplex results.

In Fig. 3 Σwr^2 is the relief parameter in a plot of mass rato q against Ω. The results are shown for 16 simplex trials in which the starting values spanned the ranges $5 \le q \le 7.5$ and $8 \le \Omega \le 13$. The minimum is steep-sided despite the artifact of many side valleys. The 2-D contour plot in Fig. 4 shows the fine agreement between final solutions with both programs. The error bars for the *WD* solution are from the rms full parameter set fit errors.

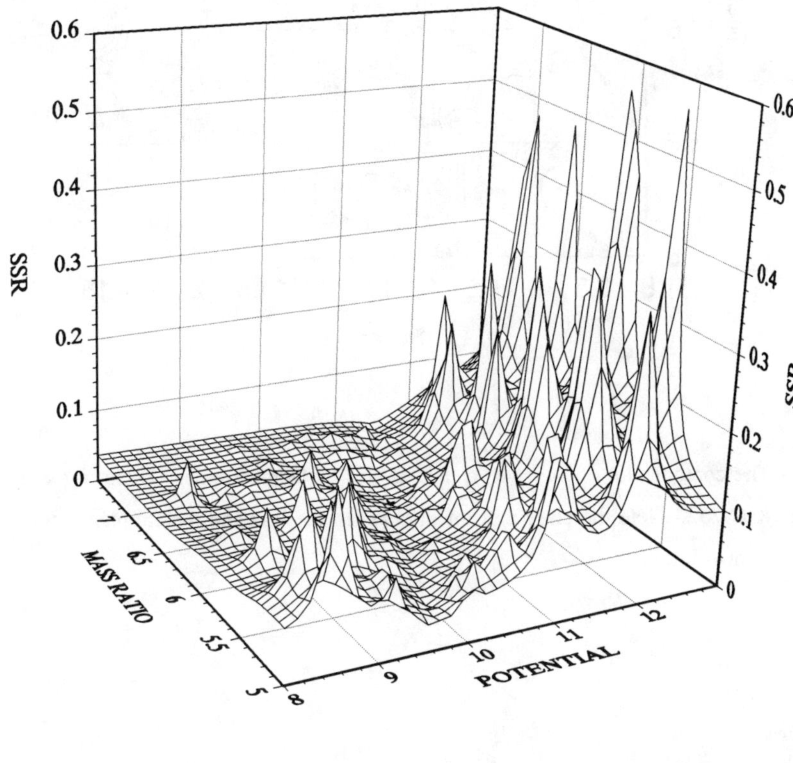

Fig. 3 - Σwr^2 *vs*. q and Ω for the contact binary H235 in the open cluster NGC 752.

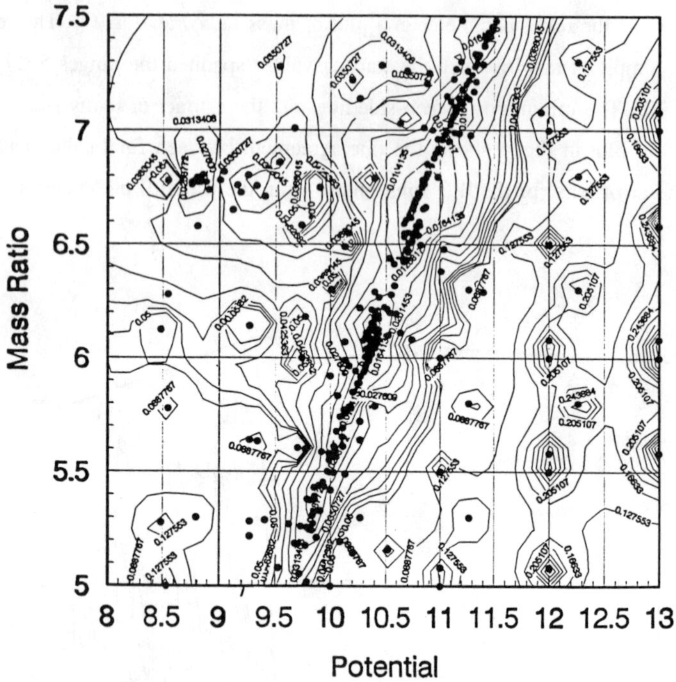

Fig. 4 - The contour plot of Σwr^2 *vs.* q and Ω for the contact binary H235 in the open cluster NGC 752. Note the close agreement between the *WD* and the simplex enhancement solutions.

Portability

We as well as many other groups have been forced to use a number of computing platforms over the past few years. The following have been used at the University of Calgary:

* Honeywell-Multics computer
* Cyber 205 supercomputer
* Myrias SPS-2 fine grain (64-processor) parallel computer
* IBM RS6000 530 and 320 server and workstation;
* Sun SPARCstation; and
* Zenith Z-386 16 MHz microcomputer, an alternative platform, not usually used because of its relative slowness (however, a Datel 486 33 MHz machine approaches within a factor of 3).

The Z-386 machine is used as a smart terminal, which can produce graphics output immediately with plotting packages.

* the differential corrections version of the code is run on a Myrias SPS-2 64-processor machine, as well as on Sun and IBM workstations;
* the Z-386 is used for routinely plotting the LC output and the *MULTICS* is used when final hard-copy is desired.

Codes of any degree of complexity cannot be expected to be used on any single platform interminably. In 1985 (Milone 1989), it was argued that a supercomputer appeared to be the best hope for the extensive computing required for multi-passband, multi-epoch light curve analyses. Today, however, the Cyber 205 (a supercomputer of the 1980s with high operating and maintenance costs) is not supported any longer at the University of Calgary. While the 205 is still the fastest single processor yet available to us, we can no longer depend on it for light curve analysis.

CRS has adapted the *WD* program for the Myrias SPS-2, a massively parallel machine, which has proved to be a very important interim platform for us. The Myrias corporation went into receivership in 1990, but the University of Calgary is committed to maintaining the present machine over a five year period. The operating system of the Myrias computer is *UNIX* based, as are most new computing platforms, and the code, now parallelized, can readily be moved to another fine-grain parallel machine. Two comparisons of computer platforms were done. The first one used a DC input with 122 data points, 6 parameters to be varied, and a grid size of 20 for each star. The results are summarized in Table 1. The second comparison used a DC input with 703 data points, 15 parameters to be varied, and a grid size of 20 for each star.

Table 1. Comparison #1 (times in seconds):

time-	**Myrias (64-proc.)**	**Sun**	**IBM**
elapsed	93	515	268
user	50*	448	129
system	2.72	7.50	0.34

*** average per processor**

Table 2. Comparison #2 (times in seconds)

time-	Myrias (60-proc)	Sun	IBM
elapsed	673	4366	3339
user	529*	4339	1491
system	11.78	1.9	0.7

*** average per processor**

Future Developments

In the immediate future, the best possible atmospheres must continue to be incorporated, and we look forward to Kurucz's further improvements (described elsewhere in this volume): to cooler stars, and increased dimensionality of the grid to include abundances, and mixing length/scale height. This is especially important to studies of binaries in clusters (*cf* Milone and Schiller 1984, 1988).

The use of mean effective temperatures as modeling parameters in light curve modeling has recently been criticized (Popper 1990 and in these proceedings), and probably there is room for still better physics in the modeling codes. Since the mean stellar flux arises from a range of physical depth that can be different from passband to passband, it could, in principal, be modeled with different color temperatures. We think this is equivalent to saying that stars are not in thermodynamic equilibrium, which is no news, and light curve codes should contain the means to deal with it. There certainly are cases even in the improved versions of the WD program, where the decoupling of T and L provides better fittings to each passband than can be obtained with a comprehensive fitting (*e.g.* in AI Phe - see Milone, Stagg and Kurucz 1991). Whether departures from LTE or not are the cause for this, it seems to us that modern empirical or quasi-empirical atmospheres are very important at this stage of the field's development; eventually it is to be hoped that all of the parameters which create the necessity for such empiricism will be fully quantifiable. As we move through and beyond millimagnitude scale modeling, this will require the best that theoretical atmosphere studies can provide in the optical passbands; it is already a problem in the IR and far UV, which arise from quite different regions of stellar atmospheres. The use of the best atmospheres is required also if absolute fluxes are to be determined, which, we argue below, will provide a valuable additional tool for understanding light curve perturbations. There is mounting evidence that non-linear limb-darkening laws obtain in at least solar type binaries in the UV; therefore, it is time for a non-linear option to light curve modeling. This suggestion is made elsewhere in these proceedings also.

The specialized used of thin and thick disks, including those in at least some of the β Lyrae systems (the *W Serpentis stars* of Plavec and his collaborators), is an area that has attracted some attention (see Antokhina and Menchenkova 1990, and references cited therein). We look forward to the opportunity to model RT Lacertae (Milone 1976, 1977) with the 'double occultation' model that can be brought to bear on it through Russell-Merrill techniques but is currently not so easy to model with the light curve synthesis codes. This 5.07^d-system of cool subgiants is a curious cross between *W Serpentis* and RS CVn-like systems. It is marginally detached or semidetached (depending on the model) and it has been modeled with spots, which, however, may not do justice to the variety of mysteries associated with it. Modeling work on CBS with circumstellar envelopes has been attempted, for example, by Pustyl'nik and Éinasto (1987). More exotic objects, such as symbiotic stars, cataclysmic variables, x-ray binaries, and perhaps B_e stars, where disks of various kinds are thought to play a role, are also being modeled by a number of groups.

The success of those modeling efforts will have to await some other forum, but the existence of such cases are a warning that codes need to be flexible.

The usefulness of a synthetic light curve and spectral analysis code is greatly increased if the code can highlight and diagnose light curve and spectral perturbations from a standard model. In the past, Milone has described the value of infrared observations and spectrophotometry in providing aspects of such a tool. As a working group from IAU Commission 25 is at work to improve IR extinction determinations and the precision of IR transformations, no more needs to be said about this at the present time, but this is where better atmospheres will prove the most valuable: to demonstrate the where and the why of the perturbations.

The use of Doppler imaging and magnetic intensification effects on high dispersion spectra provide obvious improvements for the next five years. This development thus proposes the mating of high spectral resolution to millimagnitude photometric precision and accuracy to produce highly monochromatic light curves. One area which is being special attention now is that of stellar wind collisions in very early systems (see work cited by Kallrath, and, for example, that of Bairamov, *et al.*, 1990); the effects of such phenomena on obervable phenomena is under study by Sreenivasan and colleagues at the University of Calgary; we should be able to tackle properly the analyses of the most interesting systems and render their envelope properties with precision. Temporal variation may be expected in such systems, and is prevalent in the more general class of close binaries. The O'Connell effect is widespread in close binary light curves and there probably are no true cases of a constant O'Connell effect. Therefore the future requires a dynamical modeling code where variations in perturbed region parameters are permitted to be time variable. Several participants here have modeled time-variable spot parameters. However, if a code is to be a fully fledged diagnostic tool to find the radiation mechanisms responsible for light curve anomalies, an appropriate level of precision for the calibration of spectrophotometric and broad-band IR fluxes will be necessary.

While we cannot be certain of the direction to be taken in the development of the fast computers of the 21st century, the 1990s are being organized around servers and workstations and common operating systems, like *UNIX*. There is, nevertheless, a case to be made for adapting a code which contains Doppler mapping spectroscopy and spectrophotometry with hundreds if not thousands of passbands to the fastest platform available - and that will probably mean a finely grained parallel *vector* machine. At present, since the Cyber 205 is the only vectorized platform regularly available to us, and its lifetime is uncertain, we have not proceeded to rewrite the code for easier pipeline processing as we had intended earlier; if another such platform materializes in the next few years, we plan to adapt any successor code in use at that time to it, resources permitting.

It is a pleasure to thank the following persons: R.E. Wilson for providing his updated spot modeling code in 1985 and for his abundant help and support on this project over the years; R.H. Kurucz for providing his atmospheres; D.H. Bradstreet for the *Binary-Maker* software and for observational data on TY Boo; and a number of collaborators now or formerly at the University of Calgary whose contributions have helped to advance the project: Gaston Groisman, B.J. Hrivnak, D.H. Krouse, J.C. McVean, S.J. Schiller, and B.J. Sugars. This work has been supported by grants from the Natural Sciences and Engineering Council of Canada to *EFM*, federal and provincial student employment programs, the University of Calgary Research Grants Committee in the form of research grants to *EFM* and a Postdoctoral Fellowship to *CRS* and by the Department of Physics and Astronomy.

References

Antokhina, É.A., and Menchenkova, E.V., 1990. *Sov. Astron. Lett.*, **16** (3), 227.

Bairamov, Z.T., Pilyugin, N.N., and Usov, V.V., 1990. *Sov. Astron.*, **34** (5), 502.

Carbon, D.F., and Gingerich, O.J., 1969, in *Theory and Observations of Normal Stellar Atmospheres*, ed. O.J. Gingerich (Cambridge, MA: MIT Press), p. 377.

Groisman, G., 1990. *The Shallow Contact System TY Bootis*. MSc. Thesis. (Calgary: The University of Calgary).

Kallrath, J., 1990. *MNRAS* **248**, 655.

Kallath, J., Milone, E.F., and Stagg, C.R., 1992. *ApJ* **389**, 590..

Kurucz, R.L., 1979. *ApJS* **40**, 1.

Milone, E.F., 1976. *ApJS* **31**, 93.

Milone, E.F., 1988. "Implications of the O'Connell Effect for Close Binary System Modeling", in *Critical Observations vs. Physical Models for Close Binary Stars* ['The Beijing Colloquium on Close Binary Stars'], ed. K.-C. Leung, (New York: Gordon and Breach) p. 139.

Milone, E.F., 1989. *Infrared Extinction and Standardization*. (Heidelberg: Springer-Verlag).

Milone, E.F., 1991. "Toward the Realization of Modeling Improvement Suggestions Made at the Beijing Colloquium on Close Binary Stars", in *International Colloquium on Developments in Close Binary Stars*, ed. K.-C. Leung and I.-S. Nha (Provo: BYU Press), *in press.*

Milone, E.F., and Schiller, S.J., 1984. *PASP* **96**, 791.

Milone, E.F., and Schiller, S.J., 1988. "A Progress Report on the RAO Binaries-In-Clusters Program and its Relevance for Distance Scales", in *The Extragalactic Distance Scale*, S. van den Bergh and C.J. Pritchett, eds., *A.S.P. Conference Series*, **4**, 182.

Milone, E.F., Schiller, S.J., Groisman, G., and Fry, D.J.I., 1989. "Modeling of Light Curves of Eclipsing Binaries on the Cyber 205 Supercomputer", in *Applications of Computers to Dynamical Astronomy*, J. Kovalevsky and P.K. Seidelmann, eds., (Dordrecht: Reidel), 45.

Milone, E.F., Stagg, C.R., and Kallrath, J., 1992. "The Eclipsing Binaries in NGC5466 and their Implications for System Evolution", in *Evolutionary Processes in Interacting Binary Stars*, R. Polidan, ed. (Dordrecht: Kluwer), *in press.*

Milone, E.F., Stagg, C.R., and Kurucz, R.L., 1992. *ApJS* **79**, 123.

Milone, E.F., Stagg, C.R., Sugars, B.J.A., McVean, J.R., Schilller, S.J., and Kallrath, J., 1992. "The Contact Binary H235 in the Open Cluster NGC 752", in *Evolutionary Processes in Interacting Binary Stars*, R. Polidan, ed. (Dordrecht: Kluwer), *in press.*

Milone, E.F., Wilson, R.E., and Hrivnak, B.J., 1987. *ApJ*, **319**, 325.

Pustyl'nik, I. and Éinasto, L., 1987. *Soviet Astron. Lett.*, **13** (4), 252.

Schiller, S,J. and Milone, E.F., 1987. *AJ*, **93**, 1471.

Schiller, S.J. and Milone, E.F., 1988. *AJ*, **95**, 1466.

Wade, R.A., and Rucinski, S.M., 1985. *A&A*, **60**, 471.

Wilson, R.E., 1979. *ApJ*, **234**, 1054.

Wilson, R.E., and Biermann, P., 1976. *A&A*, **48**, 349.

Wilson, R.E., and Devinney, E.J., 1971. *ApJ*, **166**, 605.

Wilson, R.E., and Sofia, S., 1976. *ApJ*, **203**, 182.

New Atmospheres for Modelling Binaries and Disks

Robert L. Kurucz
Harvard-Smithsonian Center for Astrophysics, 60 Garden Street, Cambridge, MA 02138, USA

ABSTRACT. I have used my newly calculated iron group line list together with my earlier atomic and molecular line data, 58,000,000 lines total, to compute new opacities for the temperature range 2000K to 200000K. Calculations have been completed at the San Diego Supercomputer Center for 56 temperatures, for 21 pressures, for microturbulent velocities 0, 1, 2, 4, and 8 km/s, for 3,500,000 wavelength points divided into 1221 intervals from 10 to 10000 nm, for scaled solar abundances [+1.0], [+0.5], [+0.3], [+0.2], [+0.1], [+0.0], [-0.1], [-0.2], [-0.3], [-0.5], [-1.0], [-1.5], [-2.0], [-2.5], [-3.0], [-3.5], [-4.0], [-4.5], and [-5.0]. I have rewritten my model atmosphere program to use the new line opacities, additional continuous opacities, and an approximate treatment of convective overshooting. The opacity calculation was checked by computing a new theoretical solar model that matches the observed irradiance. Thus far I have completed a grid of 7000 model atmospheres at 2 km/s for all the abundances, for the temperature range 3500K to 50000K, and for log g from 0.0 to 5.0. This grid will allow a consistent theoretical treatment of photometry from K stars to B stars. Fluxes and intensities are tabulated from .09 to 160 micrometers. Preliminary results are reported for many photometric systems. Work is underway on grids for other microturbulent velocities. Microturbulent velocity strongly affects the interpretation of Cepheid and RR Lyrae photometry. The models, fluxes, intensities, and colors are available on magnetic tape and will also be distributed on CD-ROMs.

1. Old models

I start by listing the shortcomings in my old models as I did at an earlier meeting (Kurucz 1987) but this time I can report that I have corrected most of them.

My old models (Kurucz 1979a,b) were produced as long as 18 years ago on computers that are primitive by today's standards. The number of optical depth layers was limited by small memories and slow processors. Now I can compute with many more layers and go to shallower optical depths. This greatly improves the numerical accuracy of the calculated radiation field at wavelengths that have very high or rapidly varying opacity. Fortunately, such wavelengths do not very much affect the structure of a model. There was also a limit to the number of frequencies that I could afford to compute. Now I use up to 1221 wavelength intervals from 9 to 160,000 nm. I can now use 1 nm resolution in the ultraviolet for better comparison to satellite observations.

My old models for F and G stars are systematically in error and predict color indices that are off by as much as 0.05 mag. I assumed that the error was caused by problems in the mixing length treatment of convection and by the omission of molecular line opacity in the coolest models. My theoretical model for the solar photosphere has several per cent error in the flux in the red and, of course, cannot reproduce the molecular features in the ultraviolet. Improvements in my treatment of convection, even going so far as having hot and cold streams, have reduced the error somewhat for the hotter convective models. I have also added approximate overshooting and have tried various increases in the mixing-length to scale-height ratio. I now use 1.25. However, I am now convinced that most of the error comes from missing line opacity, including missing atomic line opacity, which turns out to be significant at all effective temperatures. I hope that Nordlund and others will be able to produce models with realistic convection cells to take care of the convection physics, so now I am concentrating on the opacity. I discuss molecular and atomic opacity in the next section.

We do not know much about microturbulent velocity. It has been decreasing as a function of time as the models have improved. In the sun it is depth-dependent varying from about 0.5 to 1.8 km/s. The models assume a constant value. Twenty five years ago, I arbitrarily chose 2 km/s as a nice round number. In some stars it can be much larger or much smaller. Opacity, radiative acceleration, and model structure vary considerably with the microturbulent velocity. It may be that 1 or 1.5 km/s is a good choice for high gravity models. George Michaud tells me that the existence of some types of diffusion implies microturbulent velocities less than 100 m/s. Microturbulent velocity may vary strongly with phase and with depth in pulsating stars, thereby strongly affecting the atmospheric structure and colors. I plan to investigate this through spectrum synthesis. My new grids of models have microturbulent velocity as a parameter, so it must be specified when choosing a model.

The helium abundance is another arbitrary number. I chose 10% by number. Others use a 10% He/H ratio. I have switched to a smaller value because I think it more probable. Small errors in the helium abundance produce errors in the density, electron number, and opacity and consequently produce systematic errors in the derived stellar parameters. The "solar" metal abundances have also changed with time and are not yet final. I now use Anders and Grevesse (1989) abundances.

My old low gravity models have systematic errors because of non-LTE and sphericity effects. So do my new models.

2. New lines and new opacities

I reported on my line and opacity calculations at a NATO workshop in Trieste (Kurucz 1991). The details of my line lists and the opacities can be found in that paper. Here I will give only a brief outline.

My earlier model calculations used the distribution-function line opacity computed by Kurucz (1979ab) from the line data of Kurucz and Peytremann (1975). We had computed gf values for 1.7 million atomic lines for sequences up through nickel using scaled-Thomas-Fermi-Dirac wavefunctions and eigenvectors determined from least squares Slater parameter fits to the observed energy levels. That line list has provided the basic data and has since been combined with a list of additional lines, corrections, and deletions with the help of Barbara Bell and Terry Varner at the Center for Astrophysics. The line data are being continually, but slowly, improved. We collect all published data on gf values and include them in the line list whenever they appear to be more reliable than the current data. I have also completely recomputed Fe II (Kurucz 1981).

After the Kurucz-Peytremann calculations were published, I started work on line lists for diatomic molecules beginning with H_2, CO (Kurucz 1977), and SiO (Kurucz 1980). Next, Lucio Rossi of the Istituto Astrofisica Spaziale in Frascati, John Dragon of Los Alamos, and I computed line lists for electronic transitions of CH, NH, OH, MgH, SiH, CN, C_2, and TiO. In addition to lines between known levels, these lists include lines whose wavelengths are predicted and are not good enough for detailed spectrum comparisons but are quite adequate for statistical opacities. Work is continuing on other molecules and molecular ions, and on the vibration-rotation spectra. I also have data for terrestrial atmospheric molecules.

In 1983 I recomputed the opacities using the additional atomic and molecular data described above which totalled 17,000,000 lines. These opacities were used to produce improved empirical solar models (Avrett, Kurucz, and Loeser 1984), but were found to still not have enough lines. For example, there were several regions between 200 and 350 nm where the predicted solar intensities are several times higher than observed, say, 85% blocking instead of the 95% observed. The integrated flux error of these regions is several per cent of the total. In a flux constant theoretical model this error is balanced by a flux error in the red. The model thus predicts the wrong colors. In detailed ultraviolet spectrum calculations, half the intermediate strength and weak lines are missing. After many experiments, I determined that this discrepancy is caused by missing iron group atomic lines that go to excited configurations that have not been observed in the laboratory. Most laboratory work has been done with emission sources that cannot strongly populate these configurations. Stars, however, show these lines in absorption without difficulty. Including these additional lines produces a dramatic increase in opacity, both in the sun and in hotter stars. A stars have the same lines as the sun but more flux in the ultraviolet to block. In B stars and in O stars there are large effects from third and higher iron group ions. Envelope opacities that are used in interior and pulsation models are also strongly affected.

I was granted a large amount of computer time at the San Diego Supercomputer Center by NSF to carry out new calculations. To compute the iron group line lists I determined eigenvectors by combining least squares fits for levels that have been observed with computed Hartree-Fock integrals (scaled) for higher configurations including as many configurations as I can fit into a Cray. All configuration interactions are included. My computer programs have evolved from Cowan's (1968) programs. Transition integrals are computed with scaled-Thomas-Fermi-Dirac wavefunctions and the whole transition array is produced for each ion. The forbidden transitions can be computed as well. Radiative, Stark, and van der Waals damping constants and Lande g values are automatically produced for each line. The first nine ions of Ca through Ni produced 42,000,000 lines. I will recompute the energy levels and line lists when new analyses become available and I will make the predictions available to laboratory spectroscopists. I plan to do the heavier and lighter elements as a background project.

The models I can compute now are not valid for M stars. I eventually need line lists for the triatomic molecules, but I hope that other people will do the work before I have to learn the physics. I am working on the low temperature bands now, however, for atmospheric transmission.

In late 1988 I used the line data described above to compute new solar abundance opacity tables for use in my modelling. The calculations involved 58,000,000 lines, 3,500,000 wavelength points, 56 temperatures from 2000K to 200000K, 21 log pressures from -2 to +8, and 5 microturbulent velocities 0, 1, 2, 4, 8 km/s, and took a large amount of computer time. The opacity is tabulated both as 12-step distribution functions for intervals on the order of 1 to 10 nm, and as opacity sampling where, simply, every hundredth wavelength point in the calculation was saved. There are actually two sets of distribution functions, a higher resolution version with 1212 "little" intervals, and a lower resolution version with 328 "big" intervals. The "little" wavelength intervals are nominally 1 nm in the ultraviolet and 2 nm in the visible. The opacities were tested by computing a solar model as described below.

Since the beginning of 1990 I have been able to take tremendous advantage of the new Cray YMP at the San Diego Supercomputer Center. In a few months I finished more than I had expected to do in two years. I computed opacities ranging from 0.00001 solar to 10 times solar, enough to compute model atmospheres ranging from the oldest Population II stars to high abundance Am and Ap stars. The hardest part was transmitting the results (200 tapes) back to Cambridge over Internet, but even that usually worked quite well. The exact abundances are [+1.0], [+0.5], [+0.3], [+0.2], [+0.1], [+0.0], [-0.1], [-0.2], [-0.3], [-0.5], [-1.0], [-1.5], [-2.0], [-2.5], [-3.0], [-3.5], [-4.0], [-4.5], [-5.0], and [+0.0, no He]. The final files for each abundance require two 6250 bpi VAX backup tapes. I distribute copies of the tapes on request. I hope to produce CD-ROMs of these opacities that can be read on any workstation with a CD reader.

I am open to suggestions for computing opacities for other abundance mixes. I plan to compute Population II opacities with [+0.4] enhanced alpha process elements, some Am and Ap mixes, and C/O variations.

3. New models

I have rewritten my model atmosphere program to use the new line opacities, additional continuous opacities, and an approximate treatment of convective overshooting. The opacity calculation was checked by computing a small grid of solar models with various microturbulent velocities and mixing-length-to-scale-height ratios. I adopted a solar model shown in Figure 1 that matches the observed irradiance (Neckel and Labs 1984; Labs et al. 1987) with Vturb = 1.5 km/s and l/H = 1.25. I am confident that I have solved the missing opacity problem. I then computed (on Vaxstations) a grid of 400 solar abundance, 2 km/s models covering the effective temperature range from K stars to O stars. The models are listed in Table I. Models cooler than 9000K are convective with l/H = 1.25. The range of this grid should allow photometric calibrations consistent for both cool and hot stars. Note that since triatomic molecules are not included, the models cannot be used for M stars. I have distributed tapes of the models and the flux predicted from each model at 1221 wavelengths in the range .01 to 160 micrometers. The range is enough to treat ionization in H II regions and to calibrate the infrared.

I have computed preliminary UBV and uvby colors and bolometric corrections for the solar abundance grid described above following Buser and Kurucz (1978) and Relyea and Kurucz (1978) including the typographic correction from Lester et al (1986). The colors were normalized by finding the model in the grid that best interpolates the spectrophotometry of Vega (Hayes and Latham 1975; Tug, White, and Lockwood 1977), Figure 2, and that best matches the Balmer line profiles (Peterson 1969) and then by forcing the computed colors for that model to match the observed colors. That model has Teff = 9400K and log g = 3.95 just as in the earlier grid (Kurucz 1979a). However,

Table I. Solar abundance models, 2 km/s microturbulent velocity

	log g											
Teff	0.0	0.5	1.0	1.5	2.0	2.5	3.0	3.5	4.0	4.5	5.0	Teff
3500	X	X	X	X	X	X	X	X	Z	Z	Z	3500
3750	X	X	X	X	X	X	X	X	Z	Z	Z	3750
4000	X	X	X	X	X	X	X	X	X	X	X	4000
4250	X	X	X	X	X	X	X	X	X	X	X	4250
4500	X	X	X	X	X	X	X	X	X	X	X	4500
4750	X	X	X	X	X	X	X	X	X	X	X	4750
5000	X	X	X	X	X	X	X	X	X	X	X	5000
5250	X	X	X	X	X	X	X	X	X	X	X	5250
5500	X	X	X	X	X	X	X	X	X	X	X	5500
5750	X	X	X	X	X	X	X	X	X	X	X	5750
6000	X	X	X	X	X	X	X	X	X	X	X	6000
6250	B	X	X	X	X	X	X	X	X	X	X	6250
6500	B	X	X	X	X	X	X	X	X	X	X	6500
6750	B	X	X	X	X	X	X	X	X	X	X	6750
7000	B	X	X	X	X	X	X	X	X	X	X	7000
7250	B	X	X	X	X	X	X	X	X	X	X	7250
7500	B	X	X	X	X	X	X	X	X	X	X	7500
7750		B	X	X	X	X	X	X	X	X	X	7750
8000		B	X	X	X	X	X	X	X	X	X	8000
8250		B	X	X	X	X	X	X	X	X	X	8250
8500		B	X	X	X	X	X	X	X	X	X	8500
8750			B	X	X	X	X	X	X	X	X	8750
9000			B	X	X	X	X	X	X	X	X	9000
9250				B	X	X	X	X	X	X	X	9250
9500				B	X	X	X	X	X	X	X	9500
9750				B	X	X	X	X	X	X	X	9750
10000				B	X	X	X	X	X	X	X	10000
10500				B	X	X	X	X	X	X	X	10500
11000					B	X	X	X	X	X	X	11000
11500					B	X	X	X	X	X	X	11500
12000					B	X	X	X	X	X	X	12000
12500					B	X	X	X	X	X	X	12500
13000					B	X	X	X	X	X	X	13000
14000				B	X	X	X	X	X	X	X	14000
15000					B	X	X	X	X	X	X	15000
16000					B	X	X	X	X	X	X	16000
17000					B	X	X	X	X	X	X	17000
18000					B	X	X	X	X	X	X	18000
19000					B	X	X	X	X	X	X	19000
20000						B	X	X	X	X	X	20000
21000						B	X	X	X	X	X	21000
22000						B	X	X	X	X	X	22000
23000						B	X	X	X	X	X	23000
24000						B	X	X	X	X	X	24000
25000						B	X	X	X	X	X	25000
26000						B	X	X	X	X	X	26000
27000							B	X	X	X	X	27000
28000							B	X	X	X	X	28000
29000							B	X	X	X	X	29000
30000							B	X	X	X	X	30000
31000							B	X	X	X	X	31000
32000								B	X	X	X	32000
33000								B	X	X	X	33000
34000								B	X	X	X	34000
35000								B	X	X	X	35000
37500									B	X	X	37500
40000									B	X	X	40000
42500										B	X	42500
45000										B	X	45000
47500										B	X	47500
50000										B	X	50000

X = converged
B = blows up from radiative acceleration
Z = converged, not realistic because no water opacity

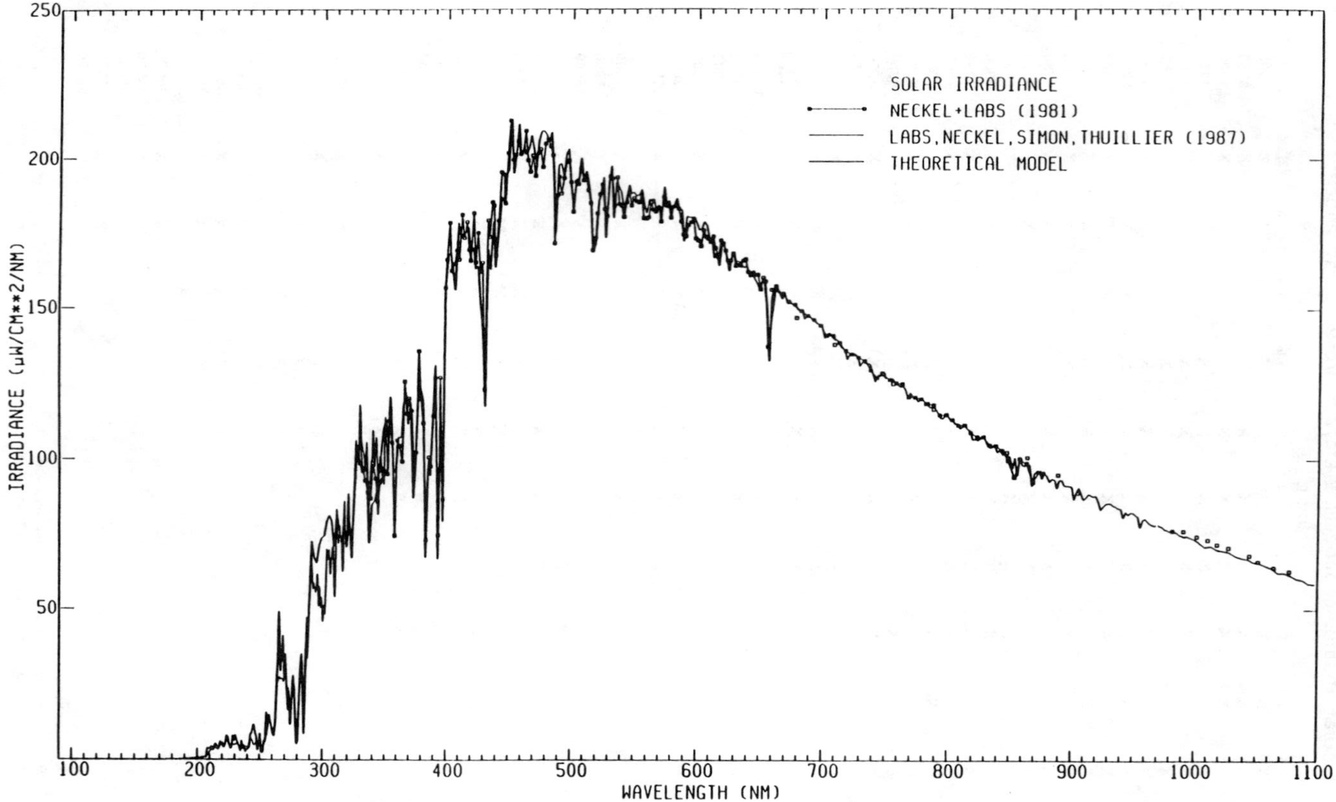

Figure 1. Predicted solar irradiance compared to observed.

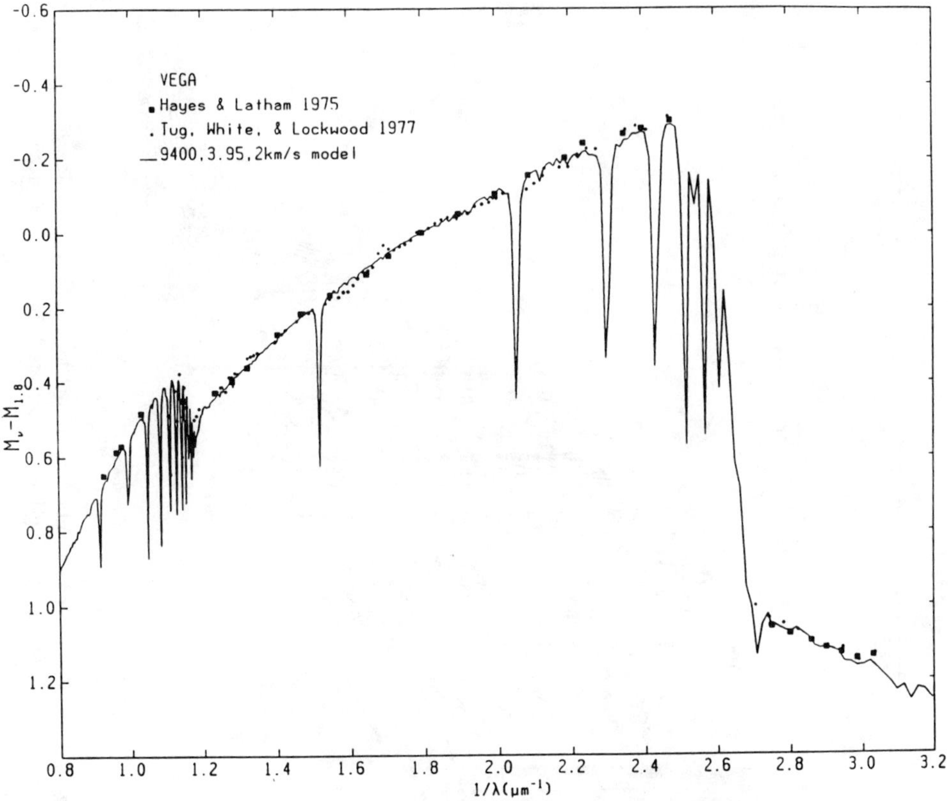

Figure 2. Observed and predicted spectrophotometry relative to 1.8 inverse micrometers (555.6 nm) for Vega. The computed slopes of the Balmer, Paschen, and Brackett continua and the Balmer discontinuity agree with observations.

remember that the helium abundance, metal abundances, line opacity, and resolution are different. Models with lower microturbulent velocity are not observationally distinguishable in the visible. This model is not a physical model for Vega because Vega does not have solar abundances (for example, see Adelman and Gulliver 1990), but even if the physical parameters are somewhat off, the temperature-pressure structure must be correct in the continuum and Balmer-line-wing-forming layers. A correct model will have a similar structure but perhaps for a somewhat different effective temperature and gravity.

I have begun to compute grids of models on the Cray using the solar abundance, 2 km/s grid for starting models. Thus far I have run approximately 7000 models for all abundances at 2 km/s. Just checking the output is a tremendous amount of work for me. I have to individually rerun mistakes and failures. I will compute the fluxes, predicted photometry, Balmer line profiles, and limb-darkening for each model. At the present

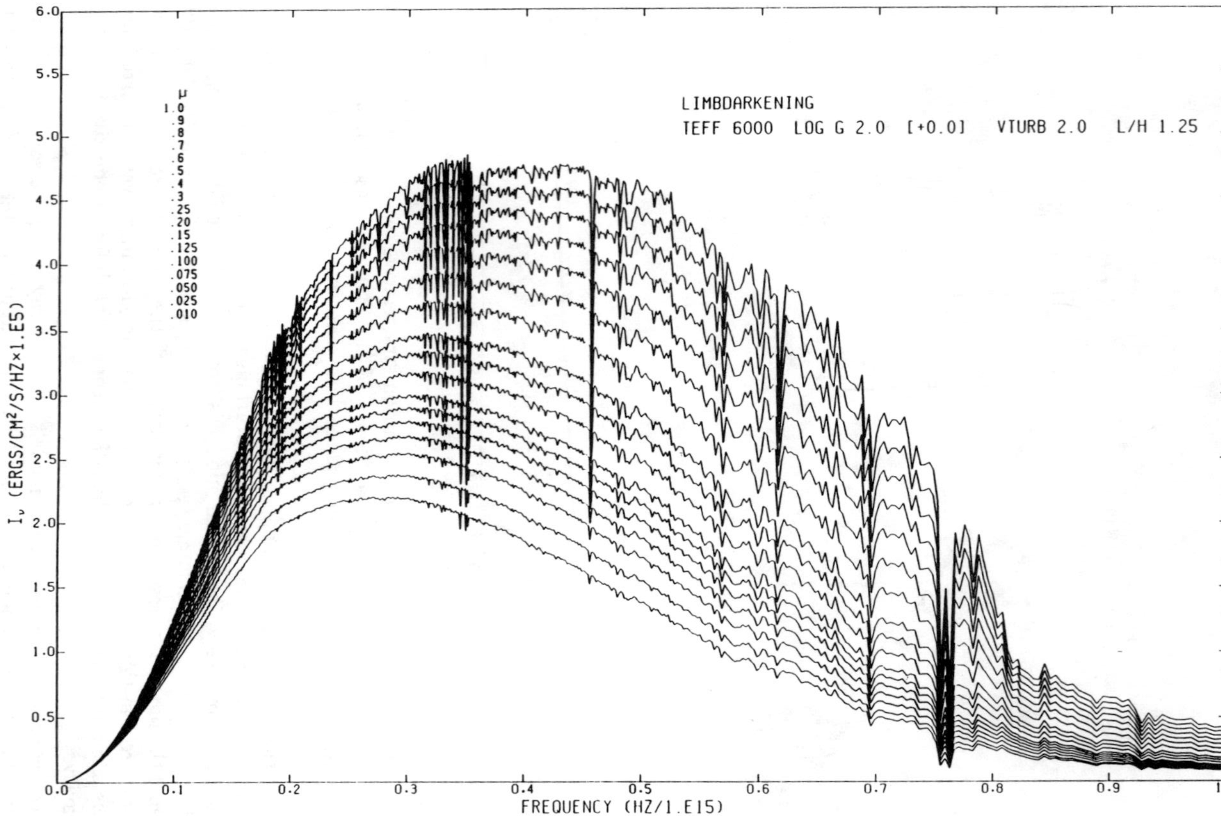

Figure 3. Limbdarkening at 17 angles for a G giant. The model has solar abundances, Teff 6000K, log g = 2, Vturb = 2 km/s, and mixing-length-to-scale-height ratio = 1.25. The 17 values of μ are listed on the figure.

present time I can supply magnetic tapes with the models for the various abundances. I will also publish the data on CD-ROMs. Most users will be able to find what they need by simple interpolation. It should be easy to write a subroutine that returns intensity for arbitrary Teff, log g, wavelength, μ, (abundance, microturbulent velocity). I expect to be able to compute a complete, full-resolution spectrum for any of the models which can be compared directly to high resolution observations, or degraded to low resolution, say, 0.1 nm.

Figure 3 shows the limbdarkening at 17 angles for a G giant. The wavelength resolution is high enough for reliable calculation of intermediate-band colors. It is also high enough to show the strongest lines and to show that the limbdarkening is not the same in the continuum and in the lines. With these models it should be possible to predict light curves that are consistent in the ultraviolet, visible, and infrared.

ACKNOWLEDGEMENTS. This work is supported in part by NASA grants NSG-7054, NAG5-824, and NAGW-1486, and has been supported in part by NSF grant AST85-18900. The most important contribution to this work is a large grant of Cray computer time at the San Diego Supercomputer Center.

REFERENCES

Adelman, S.J. and Gulliver, A.F.: 1990, *Astrophys. J.* **348**, 712.
Anders, E. and Grevesse, N.: 1989, *Geochimica et Cosmochimica Acta* **53**, 197.
Avrett, E.H., Kurucz, R.L., and Loeser, R.: 1984, *Bull. Amer. Astron. Soc.* **16**, 450.
Buser, R. and Kurucz, R.L.: 1978, *Astron. Astrophys.* **70**, 555.
Cowan, R.D.: 1968, *J. Opt. Soc. Am.* **58**, 808.
Kurucz, R.L.: 1977, *SAO Spec. Rep.* No. 374, 170 pp.
Kurucz, R.L.: 1979a, *Astrophys. J. Supp. Ser.* **40**, 1.
Kurucz, R.L.: 1979b, *Dudley Observatory Report* No. 14, ed. A.G. Davis Philip, p. 363.
Kurucz, R.L.: 1980, *Bull. Amer. Astron. Soc.* **11**, 710.
Kurucz, R.L.: 1981, *SAO Spec. Rep.* No. 390, 319 pp.
Kurucz, R.L.: 1987, *IAU Colloquium No. 95, The Second Conference on Faint Blue Stars*, eds. A.G. Davis Philip, D.S. Hayes, and J.W. Liebert, L. Davis Press, Schenectady, 129.
Kurucz, R.L.: 1991, *Stellar Atmospheres: Beyond Classical Models* ed. by L. Crivellari, I. Hubeny, and D.G. Hummer, NATO ASI Series, Kluwer, Dordrecht, 440.
Kurucz, R.L., and Peytremann, E: 1975, *SAO Spec. Rep.* No. 362, 1219 pp.
Labs, D., Neckel, H., Simon, P.C., Thuillier, G.: 1987, *Solar Phys.* **107**, 203.
Lester, J.B., Gray, R.O., and Kurucz, R.L.: 1986, *Astrophys. J. Supp. Ser.* **61**, 509.
Neckel, H. and Labs, D.: 1984, *Solar Phys.* **90**, 205.
Peterson, D.M.: 1979, Personal communication.
Relyea, L.J. and Kurucz, R.L.: 1978, *Astrophys. J. Supp. Ser.* **37**, 45.
Tug, H., White, N,M., and Lockwood, G.W.: 1977, *Astron. Astrophys.* **61**, 679.

DISCUSSION

WILSON: Can you tell us approximately how fast the subroutine you describe would run?
KURUCZ: The subroutine would read the input file and pretabulate over the desired range of temperature, gravity, and μ for each wavelength in the wavelength range of interest. There would be uniform spacing, either linear or log, and at least 100 angles. Then the subscripts for interpolation can be computed directly from the T, g, and μ values. The same linear interpolation coefficients can be used for all wavelengths. I would guess that the total cost would be equivalent to two or three exponentials per intensity.

Light Synthesis Modeling of Close Binary Stars

Albert P. Linnell
Department of Physics and Astronomy, Michigan State University, East Lansing, MI 48824, USA

ABSTRACT

Recent years have seen substantial progress in understanding residual departures of observed light curves, for close binary stars without degenerate components, from the canonical Roche model. Examples of observed departures include the O'Connell effect and "anomalies" in W Ursae Majoris systems.

The simulation of these residual effects requires a flexible program package, and involves the introduction of additional parameters, some of which may be time-dependent.

This progress has left unresolved a basic disagreement concerning the cause of W-type W UMa light curves. Although those who solve such light curves use the hot secondary model, there has been strong support for the starspot model. The latter model has not heretofore been subjected to a comparative test. A test on W Ursae Majoris data, with 89 spots, produces light curves and color curves that give an appreciably poorer representation of the observations than the hot secondary model.

Sparse observational data at 330, 250, and 220 nm from the Astronomical Netherlands Satellite have historically been interpreted to provide support for the starspot model. However, simulation of the individual light curves demonstrates inability of either the hot secondary or starspot model to represent those data. The light curve amplitudes require enhanced surface brightness near the neck region, which will be eclipsed at both minima. Thus, the problem of the W-type light curves of W Ursae Majoris remains incompletely resolved.

1. INTRODUCTION

Substantial progress has been made in recent years in understanding residual departures of observed light curves, for close binary stars without degenerate components, from the canonical Roche model. One example of a departure is the O'Connell effect, in which consecutive light curve maxima are unequally high. This effect is widely regarded as due to the presence of starspots (Hall 1976; Ramsey & Nations 1980; Linnell 1981; Yamasaki 1982; La Fauci & Rodono 1983; Catalano 1983; Davidge & Milone 1984; Poe & Eaton 1985; Milone, Wilson, & Hrivnak 1987; Budding & Zeilik 1987), although Zhou and Leung (1990) argue for a coriolis force origin. (Davidge & Milone proposed additional possible origins.) Recent studies have included starspot parameters among those subject to optimization in fitting observed light curves in RS CVn systems (Zeilik et al. 1988; Kang & Wilson 1989). A second example is starspot representation of "anomalies", such as light curve tilt during complete eclipse phases, as well as small shifts in apparent orbital phase and light curve amplitude, for W UMa systems (Linnell 1991a). An implication of these studies is that the list of parameters regularly needed to represent close binary systems must be enlarged beyond the basic Roche model parameters. The starspot parameters, especially longitudes, may be time-dependent.

Increasing sophistication in simulating photometric effects of starspots, and the consequent reduction in light residuals, accentuates the need for accuracy and detail in representing the photosphere. The nature of the photosphere in W UMa stars has been the subject of controversy. Existing light synthesis programs adopt an equipotential surface for the photosphere. Further, there usually is an unstated assumption that the photosphere is barotropic (equation of state variables functions only of gravitational plus centrifugal potential [Tassoul 1978, ch. 4]). This condition implies an isothermal photosphere. In the absence of limb darkening transit and occultation eclipses would be equally deep. Limb darkening deepens the transit eclipse and produces A-type light curves. The W-type light curves observed in W UMa stars of spectral types later than about F5, in which the deeper primary minimum corresponds to occultation eclipse, must associate with departure from the basic barotropic assumption. The nature of the departure has been a matter of contention.

A barotropic photosphere implies exact hydrostatic equilibrium over the photosphere, and thus an absence of fluid flow, either vertically or horizontally. The necessity of intercomponent circulation currents in contact systems, to transport luminosity to the secondary component (Linnell 1987, 1991a) is incompatible with this adopted condition, unless the circulation occurs well below the photosphere. No circulation model currently is available to evaluate how large this effect may be on observational data. The fact that the photosphere is not an equipotential surface (Linnell 1991a) also must affect the light synthesis model at some level.

High speed photometry, which can produce thousands of accurate multicolor observations in one night (Linnell 1980, 1982, 1985), also requires an accurate physical model to extract maximum information from the observations. It is undesirable to combine observations into normal points. Phase resolution is lost, and it becomes difficult to identify and remove a few discrepant observations. The light synthesis program should be able to process an arbitrarily large number of observations in a single run.

2. THE LIGHT SYNTHESIS PROGRAM

The analyses described herein have employed the author's light synthesis program (Linnell 1984). Parameter optimization has been described in a separate publication (Linnell 1989). This section contains a summary of the salient features of the program package.

The production of a theoretical light curve involves five consecutive programs that isolate logically distinct processing features. The programs produce cataloged output files that serve as input files to subsequent programs. Each program has an input information file in standardized format that supplies system parameters, together with other necessary control information. The programs currently run on VAX machines under DCL commands. Each program has an execution file that supplies names of necessary input files. Longer concatenated execution files link the programs together in a controlled execution sequence. The longest execution file does a complete optimization, and includes a command sequence for a differentials correction program, as well as individual programs that update the input files for the simulation sequence, as the optimization proceeds autonomously for a specified number of iterations. A description follows of the separate programs and their characteristics.

2.1. *PGA*

This program sets system geometry for the Roche model. The system to be simulated may be detached, semi-detached, or contact. Current array dimensions allow up to 49 latitude divisions, counting the two poles, and up to 2000 total division points per component. Direction cosine arrays are calculated for radii vectors to all division points, and vectors normal to the photosphere at the division points. Local gravity values are calculated at all division points. *PGA* calculates the volume of each component and lists it, with other data, in an output file.

2.2. *PGB*

This program sets an array of fiducial orbital longitudes at which light values will be calculated (in a subsequent program). For each orbital longitude, and at the assigned orbital inclination, the program calculates the bounding longitudes for segments that are visible to the observer on each latitude profile. There may be up to three such segments, on a given profile, on an overcontact system and on profiles near the neck region. The program determines outer bounding horizon points for the eclipsing star and fits parabolas to consecutive groups of three points on the outer horizon of this star. Note that a visible segment may end and another start within the outer bounding horizon on a given latitude profile, for a particular orbital inclination and longitude. These data are used to determine shadow boundaries on the eclipsed star, on each latitude profile. This is the most complex program in the simulation package.

2.3. *PGC*

This program determines the temperature and the light intensity normal to the photosphere at each grid point. It also determines first order and second order limb darkening coefficients at each grid point by fast interpolation in an external table of coefficients, tabulated as functions of temperature, wavelength, and log(gravity). Different tabulations may be invoked by changing the name of the external table in the execution control file for this program. An input list provides the program with the wavelengths of observation. The program determines the normal light intensities for this wavelength list either from an external model atmosphere tabulation or from the Planck law. The temperature at a given grid point depends on the assigned polar temperature for the component in question and the value of the gravity brightening coefficient. Irradiation by the companion also affects the local effective temperature. The program evaluates this effect by integrating over the visible part of the disk of the companion, assuming a limb darkening coefficient of 0.6 for the companion. The program applies a single bolometric albedo to the entire component. This is only a rough approximation. An output file lists grid point temperatures, both without and with irradiation effects, for examination.

The polar temperature of one component must be assigned by external information, such as a measured system color at a convenient orbital phase, and must remain fixed during the entire simulation process. The polar temperature of the other component can be adjusted by the optimization process.

2.4. *PGD*

PGD combines the outputs of *PGB* and *PGC* and performs the actual integrations over the two components of the binary system. The integration is accomplished by application of Simpson's Rule to the integrand at the grid points. The tabulated output relates to the set of fiducial orbital longitudes in *PGB*. *PGD* calculates (1) the surface area of each component, (2) the bolometric luminosity of each component, (3) the total monochromatic flux for each component at each observation wavelength, (4) the projected area of each component at each fiducial orbital longitude, (5) the monochromatic radiated power toward the observer at each observation wavelength and at each fiducial orbital longitude. The latter calculation includes the radiated power toward the observer from each component in the absence of eclipse effects, the amount of radiation subtracted by eclipse, and the net radiation toward the observer. In this way it is possible to separate the effects of light variation due to oblateness from light loss due to eclipses.

2.5. *SPT*

This program produces the actual theoretical light curve for comparison with observations. It inputs the system light values for the set of fiducial orbital longitudes and uses a nonlinear second order interpolation routine to calculate the theoretical light at each observational orbital longitude and at each wavelength of observation. Since the observations need not be stored in computer memory, it is possible to process indefinitely large sets of observational data. The program calculates the standard deviation of the observation residuals for each wavelength of observation. This feature permits a quality of fit test for empirically chosen sets of system parameters.

SPT also permits evaluation of the effect of starspots on system light. It uses the algorithm by Budding (1977) for this purpose, and can process up to 100 spots on either, or both, components. The individual spots may be cooler or hotter than the contiguous photosphere. Starspot parameters include latitude, longitude, angular radius, and temperature difference from the adjacent photosphere.

2.6. *DIFCORR*

This is the Roche model optimization program. It adopts a least squares algorithm for improving the Roche model parameters. Partial derivatives of light values with respect to the individual Roche parameters are needed at the orbital phases of observation. External programs calculate the derivatives at the fiducial orbital longitudes. *DIFCORR* uses these data to interpolate derivatives at the individual phases of observation. The program determines improved values for a specified subset, or all, of the Roche model parameters that are subject to optimization.

The observational material included in the following discussion was slightly affected by thin clouds during part of the observation period. This fact prevented use of the full optimization program. The results reported here used empirically derived parameters.

2.7 ADDITIONAL PROGRAMS

Other programs, not described in detail, do the bookkeeping to update input files with improved system parameters, and to maintain an overall listing of all system parameters as a function of iteration number. This same summary file shows the standard deviation of residuals for one wavelength.

A separate program, *DFCSPOT*, optimizes starspot parameters. A description of this program has not yet been published. It has been tested on simulated observational data. The light curve analyst would alternate between an optimization run for Roche model parameters and an optimization run for starspot parameters. Great care must be exercised to avoid overdiscussing the observational data. It is desirable to validate the analysis by repeating the optimization on synthetic data modulated by simulated observational noise. As described above, the observational data discussed herein were affected by thin clouds during part of the observing session, so it was not possible to employ this optimization program.

3. W URSAE MAJORIS LIGHT CURVES

Now consider *UBVRI* observational data for W Ursae Majoris (Linnell 1985) consisting of 15000 observations obtained on the night of March 5-6, 1984. Details of the analysis are in a recent publication (Linnell 1991a), including externally determined values for the mass ratio, and mass of one component. The polar temperature of the hotter component was adopted, based on the measured system color.

A basic controversy in analysis of light curves of W UMa stars has been the cause of W-type light curves, such as W UMa itself exhibits. It is general practice to adopt the hot secondary model to produce the observed deeper occultation eclipse. There has been widespread support (Smith 1984; Rucinski 1985; Eaton 1986) for the alternative starspot model proposed by Mullan (1975), but, to this author's knowledge, there has never been an empirical comparison of the two models' ability to represent observational data. Webbink (1977) proposed a baroclinic photosphere model for contact binaries which provides a theoretical foundation for the hot secondary model.

As the wavelength of observation shortens, the difference in eclipse depth between occultation and transit, on the hot secondary model, increases. Observations of W UMa with the Astronomical Netherlands Satellite (*ANS*) showed that this does not happen (Eaton, Wu, & Rucinski 1980),herefter EWR, and this observational result has provided support for the starspot model.

An initial problem in simulating the March 5-6 light curves is the presence of an O'Connell effect, as well as delay of maximum light to orbital phase 0.27. There are good reasons to believe these effects may arise from starspots (Linnell 1991a, 1991b, 1991c), but that the starspots that cause them are not the ones that may cause the W-type light curves. Based on this appreciation, a hot secondary combined with two starspot pairs (and both pairs are necessary) gives an excellent representation of the I, V, and B light curves. The starspots were selected in pairs, in analogy to the sun and based on the Babcock (1963) and Parker (1970, 1977) descriptions of solar magnetic activity. Latitudes of $\pm 20^\circ$ and temperature contrast of 1500K were chosen on a similar basis, so the only adjustable spot parameters were longitude and angular size. The algorithm by Budding (1977) permitted calculation of light loss for a given spot pair position. The parameters for this simulation are in Linnell (1991a). Preservation of the longitude separation of the spot pairs, while allowing the group of four spots to drift in longitude produces all of the types of light curve anomalies reported in the historical literature (Linnell 1991a). This feature supports the basic validity of the starspot representation of the O'Connell effect and related light curve anomalies.

Now turn to a representation of the V light curve on the starspot model. In this model, the average surface brightness of the more massive component is reduced by preferential presence of starspots on that component's photosphere, resulting in reduced light loss during secondary eclipse, and W-type light curves. The spots that produce the W-type light curve—call them the W-spots—must be small and evenly distributed over the photosphere. Otherwise the light variation with phase will not be smooth, or the W-type light curve will not repeat regularly even though the O'Connell effect may change both in phase and amplitude. Since the W-spots are to produce W-type light curves in *all* W-type systems, they must be distributed up to the poles to produce W-type light curves in systems like VW Cep, which has relatively shallow eclipses. Simulation of this situation requires a compromise between number of spots and computer time. In the test reported here (Linnell 1991c), a total 85 spots, in addition to the four spots for the O'Connell effect, were chosen. The same temperature contrast was selected as for the O'Connell spots. The common angular sizes for the spots was varied empirically to produce an optimum overall fit to the V light curve. The fit is not particularly good, and illustrates the requirement for a smooth distribution of small spots, since the theoretical curve shows some roughness in secondary minimum as the secondary component covers and uncovers photospheric areas that vary in average surface brightness. Based on the relative quality of light curve fit, the hot secondary model is better. Since angular radius was the only adjustable parameter to produce a fit, nothing can be done to the starspot model to improve its representation of the observations. An increase in the temperature contrast between spots and the photosphere would have negligible effect since the spots already are essentially black as compared with the photosphere. A smaller temperature contrast conflicts with temperature contrast determinations in RS CVn systems (Zeilik et al. 1988, Kang & Wilson 1989, Strassmeier 1990), as well as observation for the sun.

Tests with *DFCSPOT* show extremely small light curve sensitivity to starspot latitude. To within observational error, the same results could have been achieved with single equatorial spots rather than spot pairs. Similarly, there is very small light curve sensitivity to starspot temperature contrast from the adjacent photosphere, at the temperature contrast chosen. Thus, there remains an unresolved uniqueness problem.

4. COLOR CURVES

A fit to the individual light curves is not the only way of looking at the data. Color curves also provide important constraints. Color curves partially isolate effects of source temperature variation from geometric eclipse effects. This separation permits display of source temperature variation in a way not apparent from a study of light curves alone. The following discussion illustrates this point.

The emitted power in the observer's direction, in the wavelength interval $d\lambda$, is

$$L_\lambda d\lambda = d\lambda \int_\theta \int_\phi I_0(1 - u_{1,\lambda} - u_{2,\lambda} + u_{1,\lambda}\cos\gamma + u_{2,\lambda}\cos^2\gamma)\frac{\cos\gamma}{\cos\beta} r^2 d\phi \sin\theta d\theta, \qquad (1)$$

where γ is the angle between the local surface normal and the line of sight, β is the angle between the local surface normal and the radius vector from the component center to the mesh point, θ and ϕ are the polar coordinates, and integration is over the photospheric regions visible to the observer. The other symbols have their usual meanings (Linnell 1984). The integration limits depend on orbital inclination and vary with orbital longitude. The emergent intensity normal to the local photosphere is related to the local emergent flux, and the flux is specified by the effective temperature, as affected by both gravity brightening and mutual irradiation.

If the photosphere is isothermal, $I_{0,\lambda}$ becomes independent of position on the photosphere and may be taken outside the integral signs. If limb darkening is independent of wavelength, equation 1 becomes

$$L_\lambda = I_{0,\lambda} A(\psi), \qquad (2)$$

where $A(\psi)$ is a quantity that varies with orbital longitude but is independent of wavelength. Consequently, a color, representing a function of the ratio of received flux at two wavelengths, becomes independent of orbital longitude because $A(\psi)$ cancels.

On the starspot model, the underlying photosphere is assumed barotropic. Consequently, the photosphere is isothermal. Gravity brightening and mutual irradiation slightly modify this result, but by a small amount. Thus, individual wavelength light curves show the full geometric effects of eclipse, while, in contrast, a color curve shows nearly complete independence of orbital phase. The color curves thus permit a sensitive test against observation that is not directly apparent with the individual light curves. The effects of wavelength-dependent limb darkening can be gauged by a simulation with wavelength-independent limb darkening, followed by a simulation with wavelength-dependent limb darkening.

Application of this theory to W UMa is complicated by the presence of an astrometric companion that must be included in the photometer diaphragm. The companion has a major effect on the *V-I* color curve, making that curve less suitable for a comparison of theory and observation. The light contribution of the companion is smaller in *B-V*, although still detectable. The *B-V* theoretical color curve, including the effects of third light and wavelength-dependent limb darkening, but without starspots to produce a W-type light curve, shows much too small amplitude at primary minimum. Addition of the W-spots increases the color amplitude slightly, but by an amount too small to fit the observational data (Linnell 1991c). The hot secondary model gives a much better fit. Thus, a simulation of the starspot model indicates that it fails to represent the observational data in the *I*, *V*, and *B* bands as well as the hot secondary model does.

5. THE UV PROBLEM

The *ANS* data at 220 and 250 nm are in conflict with the hot secondary model. But the *U* light curve already indicates a complication (Linnell 1991a). The theoretical curve in *U*, based on the parameters that provide an excellent fit in *I*, *V*, and *B*, is too shallow at secondary minimum. This indicates a uv excess associated with the larger component that increases its surface brightness in the *U* band. The complication becomes worse at 250 and 220 nm. In the original report on the *ANS* data, EWR reasonably interpreted the near equality of primary and secondary minimum depths as evidence against the hot secondary model, and, therefore, as evidence favoring the starspot model.

EWR studied color curves rather than the individual light curves. Plots of the 220 and 250 nm data show that the light curves become marginally A-type at those wavelengths. Starspots, chosen to produce a W-type light curve in the *V* band, cannot produce an A-type light curve at 220 nm. A uv excess on the larger component is a possible escape and provides a connection with the anomalous *U* light curve. A promising idea for a physical source of the excess (Linnell 1991c) comes from the study of faculae in the far uv by Lean (1989, 1991). She finds that the brightness contrast of faculae, in the uv, increases by two orders of magnitude as compared with 500 nm.

A quick test of the idea of enhanced surface brightness on the primary component follows by incrementing its temperature grid by different amounts for the *U* data, the 250 nm data and the 220 nm data. A 25K increase in the polar temperature of the primary component, on the hot secondary model, with corresponding increases at other grid points of that component, gives a good fit to the *U* data, as well as the 330 nm *ANS* observations. The same model, but at 250 nm, indicates a serious problem. The theoretical light curve is too shallow at both minima, by a fairly large amount (0.07 light units). Increasing the temperature of the primary component to produce a light curve that is marginally A-type makes primary minimum shallower, because the secondary component, eclipsed at primary minimum, then contributes a smaller fraction of the system light. No change, using the standard algorithm for assigning effective temperature at the grid points, can correct this problem. The larger light curve amplitude at *both* eclipses indicates an enhanced surface brightness toward the neck region, which is eclipsed at both primary and secondary minimum. The light synthesis program has facilities to simulate this effect, by adding a temperature increment to each mesh point on the inner face of each component. The effective temperature, T, at a given mesh point is given by the expression

$$T = T0 + C1 \sin\theta^{C2} \cos\phi^{C3} T0, \tag{3}$$

where $T0$ is the original mesh point effective temperature of the component in question, $C1, C2, C3$ are disposable constants, and θ, ϕ are polar coordinates.

The *ANS* data are too sparse to make any optimization attempt worthwhile. The observed 220 and 250 nm light curves are clearly asymmetric. A rough fit to the observations is in Linnell (1991b). It should be noted that the discrepancy between theory and observation is not a result of inaccurate choice of effective temperature at the grid points. The empirically selected grid temperatures for the 220 and 250 nm data do not produce a fit in *I*, *V*, or *B*. The slight temperature change to fit the *U* data, and the 330 nm data, is only an artifice. Apparently, non-thermal effects become important in the *U* band, and dominate at shorter wavelengths. It is unknown whether the 250 and 220 nm light curves represent permanent or transient features of W Ursae Majoris, or whether these light curves characterize W-type systems in general. It further is unknown whether these effects represent a new photospheric or chromospheric "anomaly" in the uv, or if there is a basic connection to the process of intercomponent energy transfer.

6. CONCLUSIONS

Algorithms to simulate starspots have greatly improved the ability of light synthesis programs to represent "anomalies" in close or contact binary light curves. This ability has enhanced the need to resolve the issue of the cause of W-type W Ursae Majoris light curves. A test of the starspot hypothesis shows that its representation of *I*, *V*, and *B* light curves for W UMa itself is inferior to the hot secondary model. However, neither model explains the 220 and 250 nm light curves obtained by the *ANS* satellite. Those eclipse amplitudes require enhanced surface brightness in the neck region that can be eclipsed at both eclipses. The extra brightness mat be of non-thermal origin.

REFERENCES

1. Babcock, H.W. 1963, Ann.Rev.Astr.Ap., 1, 41
2. Budding, E. 1977, Ap&SS, 48, 207
3. Budding, E., & Zeilik, M. 1987, ApJ, 319, 827
4. Catalano, S. 1983, in IAU Colloquium No. 71, Activity in Red-Dwarf Stars, eds., P.B. Byrne & M. Rodono, (Dordrecht:D.Reidel), p. 343
5. Davidge, T.J., & Milone, E.F. 1984, ApJS, 55, 571
6. Eaton, J.A. 1986, Acta Astr., 36, 79
7. Eaton, J.A., Wu, C.-C., & Rucinski, S.M. 1980, ApJ, 239, 919
8. Hall, D.S. 1976, in IAU Colloquium 29. Multiple Periodic Variable Stars, ed. W.S. Fitch (Dordrecht:D.Reidel), p. 287
9. Kang, Y.W., & Wilson, R.E. 1989, AJ, 97, 848
10. La Fauci, G., & Rodono, M. 1983, in IAU Colloquium No. 71, Activity in Red-Dwarf Stars, eds., P.B. Byrne & M. Rodono, (Dordrecht:D Reidel), p.185
11. Lean, J. 1989, Science, 244, 197
12. Lean, J. 1991, preprint
13. Linnell, A.P. 1980, PASP, 92, 202
14. Linnell, A.P. 1981, BAAS, 13, 833
15. Linnell, A.P. 1982, ApJS, 50, 85
16. Linnell, A.P. 1984, ApJS, 54, 17
17. Linnell, A.P. 1985, ApJS, 57, 611
18. Linnell, A.P. 1987, ApJ, 316, 389
19. Linnell, A.P. 1989, Space Sci. Rev., 50, 269
20. Linnell, A.P. 1991a, ApJ, 374, 307
21. Linnell, A.P. 1991b, ApJ, 379, 338
22. Linnell, A.P. 1991c, ApJ, 383, Dec. 10 issue, in press
23. Milone, E., Wilson, R.E., & Hrivnak, B.J. 1987, ApJ, 319, 325
24. Mullan, D.J. 1975, ApJ, 198, 563
25. Parker, E.N. 1970, Ann.Rev.Astr.Ap., 8, 1
26. Parker, E.N. 1977, Ann.Rev.Astr.Ap., 15, 45

27. Poe, C.H., & Eaton, J.A. 1985, ApJ, 289, 644

28. Ramsey, L.W., & Nations, J.L. 1980, ApJ, 239, L121

29. Rucinski, S.M., 1985, in Interacting Binaries, eds., P.P. Eggleton & J.E.Pringle, (Dordrecht:D.Reidel), p. 13

30. Smith, R.C. 1984, Quart. J. R. A. S., 25, 405

31. Strassmeier, K.G. 1990, ApJ, 348, 682

32. Tassoul, J.-L. 1978, Theory of Rotating Stars (Princeton: Princeton Univ. Press)

33. Webbink, R.F. 1977, ApJ, 215, 851

34. Yamasaki, A. 1982, Ap&SS, 85, 43

35. Zeilik, M., DeBlasi, C., Rhodes, M., & Budding, E. 1988, ApJ, 332, 293

36. Zhou, D.-Q., & Leung, K.-C. 1990, ApJ, 355, 271

Current Status of the EBOP Code

Paul B. Etzel
Astronomy Department, San Diego State University,
San Diego, CA 92182-0334, USA

Abstract

The efficient light-curve synthesis computer code EBOP is described. The model is purposely tied to light-curve defined parameters rather than "astrophysical" ones. The major applications of the model during the past 10 years have been to well-detached eclipsing systems where proximity effects are minimal. Such systems provide well-defined stellar parameters because of their simplicity. Astrophysical parameters can be inferred from empirically calibrated surface fluxes and standardized colors of the components. Improvements to the code have been implemented and others are planned, largely as a result of continued use and suggestions by the astronomical community.

Model Introduction and Code Summary

EBOP is simply an acronym for Eclipsing Binary Orbit Program, which is a very generic name. The synthesis code is written in FORTRAN with ample embedded comments; it is accompanied by a 30 page "User's Guide" (Etzel 1980). The model itself is frequently referred to as the "Nelson-Davis-Etzel" or NDE model, which follows the historical development of the model, not the computer code. The code was written to solve uncomplicated light curves for simple detached systems. As a result of this restriction, EBOP is very fast. For example, on an Intel 386/387 based computer, a series of four iterations on seven variables may be made in just two minutes on a light curve consisting of 1200 observations.

Nelson & Davis (1972) defined the original model in terms of limb-darkened, spherical stars. The primary star is assumed to be the hotter component, which is covered at the deeper eclipse. Crucial to the model definition is the use of the surface-brightness ratio, J_s/J_p, rather than stellar surface temperatures, to model the stellar luminosities. The ratio J_s/J_p is tied more directly to the ratio of the depths of the eclipses (i.e., the data), whereas temperatures must be coupled to surface fluxes via blackbody or stellar atmospheres theory. The apparent stellar luminosities of each component (L_p, L_s) are determined by integrating over the projected stellar disks after imposing a linear limb-darkening law to a series of concentric annuli on each component. The angle γ between the surface normal and the line of sight is evenly spaced across the stellar surfaces from 0° to 90°. The visible contribution from each component is integrated over all possible combinations of eclipses by a set of closed-form, analytical expressions for the areas of rings, disks, and sectors. An integration ring size, $\Delta\gamma$, of 5° (only 18 points) results in a computational accuracy of better that 0.0001 in luminosity units over a wide range of eclipse geometries, stellar radii, and limb-darkening cases. Eccentric orbit generalization is allowed for. Nelson & Davis chose a sequential optimization technique as the basis for their computer code to determine the model parameters for a given observed light curve. The user had to be wary of local minima in multi-parameter space.

Etzel (1975) made significant changes to the model definition in regard to the proximity effects of ellipticity and reflection, the treatment of third light, and the scaling of the computed light curve to the observations. (Corrections to the original paper by Nelson & Davis were also made; they are summarized in Appendix A of Etzel 1981). The underlying philosophy was to restrict the model to simple systems where variations in proximity effects should be minimal during stellar eclipses when the projected stellar disks appear to be essentially circular. Proximity effects are treated as simple phase-dependent perturbations to the luminosity of each component and are not actually involved in the eclipse calculations.

The basic input parameters of unperturbed, fractional stellar radii, r_p and r_s, and mass ratio, $q=\mathfrak{M}_s/\mathfrak{M}_p$, are used to compute the biaxial dimensions of each star assuming an n=5 polytrope following the formalism of Chandrasekhar (1933). The semiminor axes of the stars (b_p, b_s) are then employed as the radii of the components for eclipse calculations. (It is also possible to impose sphericity on the stars for direct comparison to the Russell-Merrill (1952) model if so desired). The ellipticity of each component (ε_p, ε_s) and the effects of limb- and gravity-darkening (u_p, u_s and y_p, y_s) are introduced using the expressions given by Binnendijk (1960). The ellipticity perturbations are generally adequate for values of ε less than about 0.04, which typically corresponds to a mean component radius of less than about 0.25 for values of q near unity.

The effects of reflection are treated essentially the same as by Binnendijk for uniformly illuminated hemispheres; although allowance is made for the eclipse of reflected light. The reflection treatment is based upon a simple bolometric phase law, but the actual contribution from each component (S_p, S_s) is expressed as a fractional monochromatic luminosity contribution. Reflection may be included as an adjustable parameter to assess its importance for a given system; if it is found to be significant (such as for a classical Algol), then another more advanced model should be used for the final solutions. Finally, a fairly standard differential corrections least-squares algorithm was employed to adjust the model parameters to fit the observed data; the resultant computer code was named EBOP.

The NDE model and EBOP code have undergone further refinements in conjunction with its continued use. Popper & Etzel (1981) applied EBOP to seven detached systems, all of which have components within the main-sequence band. Two important revisions were made to the model at that time. The first was the choice of the fractional radius of the primary component, r_p, and the ratio of the radii, $k=r_s/r_p$, rather than the individual stellar radii as parameters. This choice was made because k is closely tied to the ratio of the stellar luminosities, which is a parameter that can be constrained spectroscopically from the ratio of absorption-line strengths. The second was the choice of $e\cos\omega$ and $e\sin\omega$ as parameters to account for orbital eccentricity rather than e and ω alone because the former pair is more directly related to the light curve than the latter. To the first order, the displacement of the minima from exactly one-half phase is directly related to $e\cos\omega$, whereas the difference in phase widths of the two minima is related to $e\sin\omega$. For systems with very low projected orbital eccentricity, $e\cos\omega$ can be determined to a high degree of precision (within a few ten thousandths). Popper and Etzel found convincing evidence of such small orbital eccentricities for four systems in their study, whereas a casual examination of those light curves would have suggested circular orbits. Conversely, $e\sin\omega$ is a relatively insensitive parameter that should be fixed at zero (after judicious testing) for many cases where $e< 0.05$ or so.

The basic insensitivity of $e\sin\omega$, for systems with low to moderate orbital eccentricity, is a strong argument against defining e and ω as variables for light curve solutions. Such a choice can sometimes produce indeterminate results such as confusion about the nature of an eclipse being a transit or occultation, or, at the very least, a quadrant error in ω. It is noteworthy that for spectroscopic orbit solutions the quantity $e\sin\omega$ is better determined than $e\cos\omega$ because the former is more sensitive to the temporal departure of the radial velocity curve from a sinusoid. The difficulties of determining spectroscopic values of $e<0.05$ (Lucy and Sweeney 1971) still persist even with today's high signal-to-noise detectors. Examples of eclipsing systems with moderately high orbital eccentricity that give consistent photometric and spectroscopic results are V1134 Cygni (Popper & Etzel) and α Coronae Borealis (Tomkin & Popper 1986), which have eccentricities of 0.557 and 0.370, respectively.

The reader is referred to Etzel (1981) for a concise synopsis of model parameters and the analytical expressions used in the EBOP code. Various users of EBOP have since suggested or made improvements to the model or the distributed code. Forthcoming and planned improvements will be discussed below.

Code Usage

The user is given complete control over which parameters are allowed to be variable or held fixed. Data input has been made as flexible as possible. The light curve is input as a separate file of either normal points or individual observations; currently, only a single filter file is input per run. Combinations of phase or Heliocentric Julian Date and magnitudes (sense of variable minus comparison or *vice versa*) or fluxes may be input along with an optional weight (e.g., number of observations per normal). At the present time, the supplied ephemeris information of period, P, and time of mid-primary minimum, T, are only used to compute phases at input. A phase-shift parameter, $\Delta\theta$, may be solved in order to remove any ephemeris error. Observations or normal points may also be additionally weighted by various observed-luminosity factors (e.g., $1/\ell$ and $1/\ell^2$). At execution time, the average weight is normalized to unity. An associated parameter file contains records that control the solution or series of solutions for a particular run of EBOP. Parameter values and their status (whether fixed or variable) and solution control flags are initialized in this file. The data format for the observations may also be input at execution time in the parameter file. Output for every iteration during a run (series of iterations) provides for each parameter its current value, solved correction, and standard error. An output tabulation is provided for the comparison of the calculated light curve to the observed data, which includes the magnitude and luminosity residuals, luminosity contributions of each component, and the light lost from eclipses. Full-page printer plots of magnitude residuals through each eclipse and the two maxima are produced. Magnitude residuals are preferred for these plots over luminosity residuals since for most photometry the errors of observation are approximately constant for magnitudes (Popper & Etzel 1981). These plots are of sufficiently high resolution to provide a very useful diagnostic to aid in the selection or modification of model parameters. Systematic errors make themselves very evident in the residual plots; the user learns to identify certain residual patterns at various phases with inadequacies in the values of their associated parameters (i.e., astronomy is done). Auxiliary output data files can be produced for use with a specialized graphics package that may be resident on the user's host machine to plot theoretical light curves, residual plots, or for the calculation of colors when multicolor solutions are done. In many respects, graphical diagnostics provide the best methods to assess light-curve solutions.

Figure 1 illustrates the Strömgren y-filter light curve by Thurman (1990) of the well-detached system EG Serpentis as observed at San Diego State University's Mount Laguna Observatory. The system is composed of two very similar early A-type stars with an orbital period of 9.97 days! (Thurman and his assistants used both the 0.4 and 0.6 m reflectors to obtain double coverage of several eclipses to insure adequate coverage of the light curve). The relatively narrow phase widths of the eclipses, combined with their nearly equal depths of about 0.7 magnitudes, suggests nearly central eclipses by two

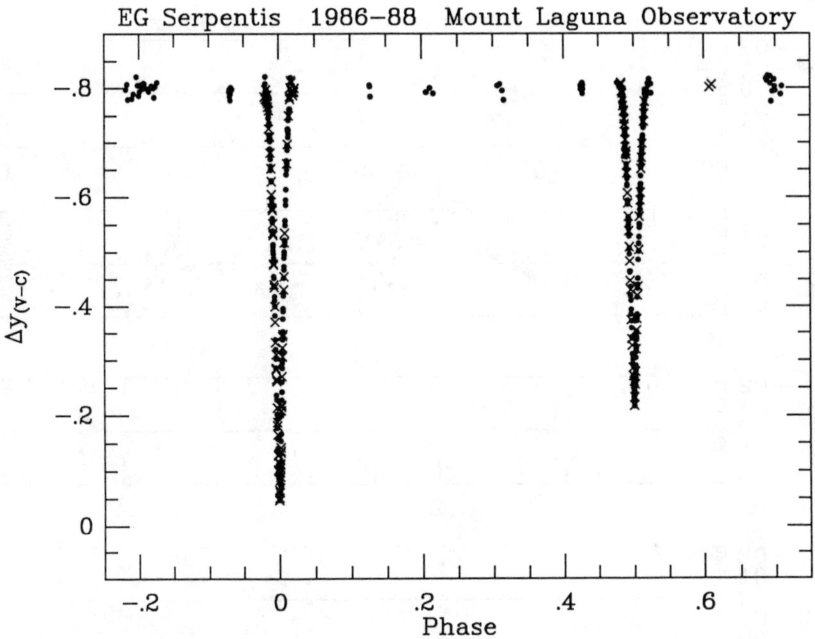

Figure 1 - The *y*-filter light curve of EG Serpentis as obtained by Thurman (1990) at Mount Laguna Observatory. Observations obtained with the 0.6 *m* reflector are plotted as dots and those with the 0.4 *m* are plotted as "X"s. The system is composed of two well separated early A-type stars. The 4-color light curves were solved with EBOP.

similar stars of small relative radii. EBOP provided solution parameters of r_p=0.056, k=0.93, and i=89.°50 from all four filters. In the *y*-filter, J_s/J_p= 0.846 was determined. All four filters verify the existence of a remarkably small orbital eccentricity (about 0.0002 in terms of $e\cos\omega$).

Figure 2 illustrates the usefulness of residual plots through the minima and maxima to judge the goodness of fit. The comparison of such plots from run to run are very helpful in assessing the sensitivity of the solution to k as was illustrated by Popper and Etzel (1981) for partially eclipsing systems where $k\cong 1$. Popper (1984) has addressed the general question of the uniqueness of solution and the sensitivity of parameters as related to light curve form and data quality. The general conclusion is that one cannot put too much faith in the formal errors provided by least squares when inferring solution uniqueness and convergence.

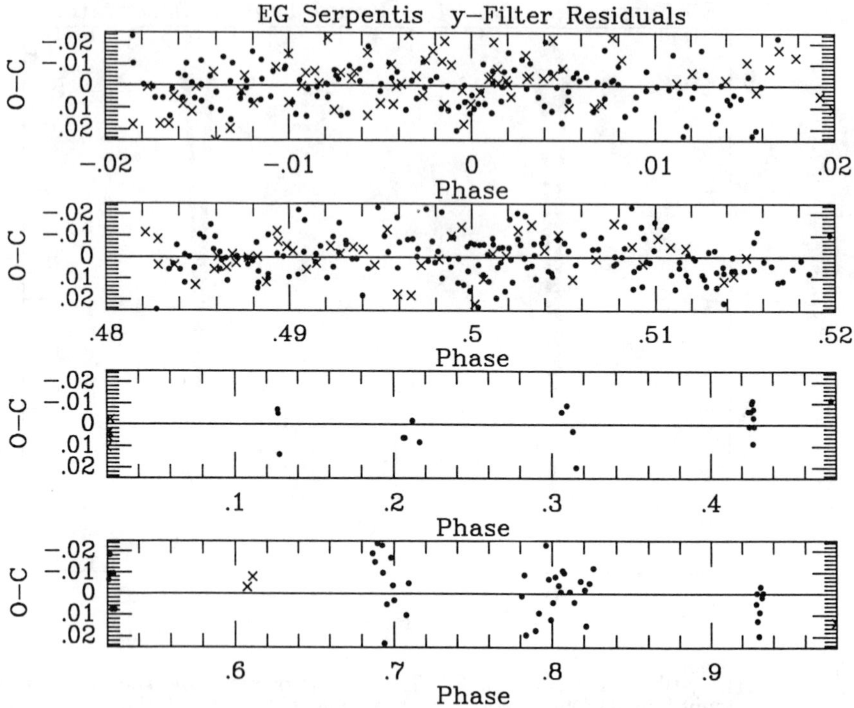

Figure 2 - The magnitude residuals from the EBOP solution by Thurman (1990) to his *y*-filter observations of EG Ser. Plotted symbols have the same meaning as in Figure 1. Few significant systematic residuals are evident in either the primary or secondary eclipses (top two panels) or the maxima (bottom two panels), especially in phase ranges covered on more than one night.

Figure 3 provides visual evidence for the consistency of the radiative parameters of the multicolor light curve solutions by plotting the observed *(u-b)* observations and the theoretical color curve as predicted by EBOP. The color changes in the two eclipses are consistent with the hotter star being eclipsed at primary eclipse. The overall astrophysical interpretation is consistent with Popper's (1986) mass determinations for the components. A full account of the work on EG Ser will be published elsewhere.

Other Representative Applications of EBOP

The majority of published applications of EBOP have dealt with the determination of fundamental stellar parameters as tests of basic stellar evolution theory. In this regard, the

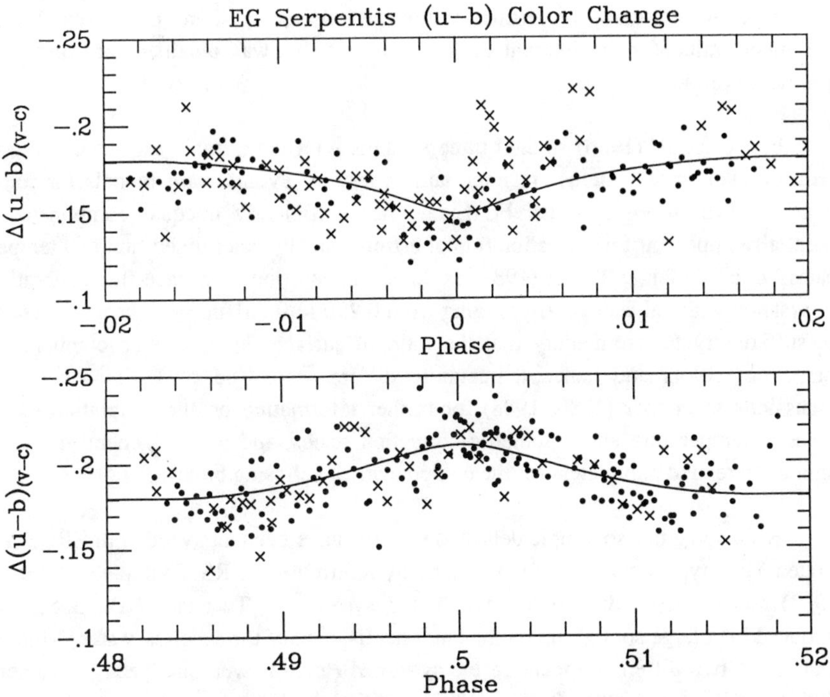

Figure 3 - Observed and computed *(u-b)* color for EG Ser as determined by Thurman (1990). The plotted symbols for the observations have the same meaning as in Figure 1. The theoretical curve was generated from the EBOP solutions for the *u* and *b* filter light curves. The color variations are due to temperature differences between the two components.

use of well detached systems is of paramount importance because of their lack of proximity complications both photometrically and spectroscopically (Popper 1980). A few representative applications are now mentioned.

Popper's (1982) photometric and spectroscopic study of the B-type, apsidal-motion binary DI Herculis (P=10.6 days, e=0.489) illustrates the use of the ratio of absorption line strengths for the two components to place constraints on the value of k for this partially eclipsing system. He also discusses the validity, or lack of it, in the use of least-squares in determining the uncertainties of both spectroscopic and photometric orbital quantities. DI Her has an apparently anomalously low apsidal motion rate, which has been discussed in the context of tests of the theory of general relativity (see Reisenberger & Guinan 1989 and references therein). Andersen et al. (1985) also used absorption-line strengths to constrain k in their study of the apsidal-motion, B-type binary V760 Scorpii (P=1.73 days,

e=0.0265). In this case, the apsidal motion period is very short (U=40 years). A well determined value for the internal density constant, k_2, was obtained for comparison to evolutionary models.

Popper et al. (1986) studied three systems with lower main-sequence components. Such relatively rare systems are important as tests of evolutionary models for solar-type stars. One of those systems, FL Lyrae, has sufficiently unequal components to be potentially important for the calibration of astrophysically determined flux and temperature scales. Lacy, Frueh & Turner (1987) explored the question of surface-flux calibrations for six systems with values of J_s/J_p ranging from 0.290 to 0.761 in the V band. These values are sufficiently far from unity that the ratio of surface fluxes can be compared to the component colors and hence temperature differences. The reader is referred to the discussions in Popper (1980, 1988) for further information on the calibration of surface fluxes to temperature and bolometric correction scales, and for the general discussion of luminosity related parameters for the components of eclipsing binaries.

Even some not-so-simple detached systems have been analyzed with EBOP, such as spotted, late-type stars, which are commonly referred to as RS CVn stars. Coates et al. (1983) modelled the distortion wave in the system CF Tucanae (HD 5305) as being produced by a large spot group on the outer hemisphere of the secondary star. Light curves obtained at two different epochs of the system Z Herculis were analyzed by Tümer et al. (1984) and Popper (1988). Popper (1990) rectified the light curve of the totally eclipsing, spotted system RW Ursae Majoris; the distortion wave was adequately modelled by a simple sine wave. He also discusses the interpretation of surface fluxes complicated by star spots and ultraviolet excesses, and their effects upon luminosity-related parameters; the fundamental properties of RS CVn-like stars are also discussed in relation to stellar evolution theory.

The evolution of stars even further from the main sequence has been explored by Andersen et al. (1991) in their study of TZ Fornacis, which is a 76 day period, detached system containing G8III and F7IV stars that have apparently evolved independently because of their wide separation ($\langle r \rangle \cong 0.05$ even at this advanced stage of evolution). Systems containing simply evolved, red-giant components are indeed rare. The masses have been determined to better than 2%; the resultant comparison to stellar evolution models has allowed the general question of convective overshooting to be explored.

The importance of further detailed study of detached systems in the context of stellar atmospheres and stellar evolution theory cannot be over emphasized. Specific uses of, or problems related to, detached eclipsing systems mentioned in the literature are: 1) there are very few uncomplicated systems known with O- or early B-type main-sequence components; 2) even though late G-, K-, and M-type dwarfs are very common, the small radii of these stars makes the probability of detecting an eclipsing binary with large separation between the components low (if they are not well separated, they may be complicated by enhanced spot activity); 3) tests of limb-darkening laws should be possible in well observed cases; 4) surface-flux calibrations and luminosity parameters need to be better established over many locations on the H-R diagram; 5) continued apsidal motion tests of stellar structure; 6) tests of the agreement between isochronal ages of binary stars of known mass in open clusters and their cluster turn-off ages; 7) the use of systems with small non-zero orbital eccentricities as possible indicators of Jovian-like planets or as tests of orbital synchronization and stellar rotation theories; 8) the chance detection of tertiary eclipses by coplanar, Jovian-like planets; and 9) the calibration of extragalactic distance indicators through the use of well understood binaries in Local Group galaxies.

Planned and Possible Improvements

Even before this workshop, I had been planning a series of revisions to the NDE model and the EBOP code. I was pleased to see the contribution by Giménez and Díaz-Cordovés (1992) because they have made some significant improvements to EBOP that I will gladly incorporate into the distributed code. Briefly, their changes are: 1) minor corrections to Binnendijk's (1960) ellipticity treatment; 2) the inclusion of two non-linear limb-darkening laws; and 3) the treatment of variable periastron position for systems with rapid apsidal motion rates.

Other possible changes that I am contemplating are: 1) removal of the phase shift parameter, $\Delta\theta$, and incorporation of the ephemeris quantities P and T as solution variables; 2) integration of apsidal motion variables and orbital eccentricity in such a way as to still allow solution of $e\cos\omega$ and $e\sin\omega$ as variables when apsidal motion is absent; 3) allowing the difference in limb-darkening between the two components to be a variable (currently one can impose an equal limb- and/or gravity-darkening condition for the two components); and 4) permitting simultaneous multicolor solutions in a limited capacity. Users should feel free to send me suggestions or their actual code changes for inclusion into the distributed code.

All efforts will be made to insure continued ease of use and cross-vendor compatibility of the code. Vendor specific graphics options will be avoided; advanced users can add such options themselves. Output data sets will still be supported so that commercially available graphics packages can be used to prepare plots such as those shown in Figures 1-3 here. The "EBOP User's Guide" will also be updated to reflect recent improvements. The primary distribution site will remain at the author's home institution. EBOP will remain a tool for uncomplicated eclipsing systems. No attempts will be made to extend the model to work on more complicated systems; other more generalized models should be used for such systems (e.g., Wilson & Devinney 1971, Wilson 1979, 1990).

Acknowledgements

I would like to thank those astronomers who have used EBOP as a scientific tool for making my investment of effort so worthwhile. Suggestions by and discussion with Daniel Popper, Edward Olson, Burt Nelson, Claud Lacy, Robert Koch, Alvaro Giménez, Walt Davis, and Johannes Andersen have proven very useful in the growth of EBOP over the years. This work was partially supported by a grant from the National Science Foundation (AST-8822790). I also thank the American Astronomical Society for a supplemental travel grant to attend this IAU sponsored meeting.

References

Andersen, J., Clausen, J. V., Nordström, B., & Popper, D. M. 1985, A&A, 151, 329

Andersen, J., Clausen, J. V., Nordström, B., Tomkin, J., & Mayor, M. 1991, A&A, 246, 99

Binnendijk, L. 1960, Properties of Double Stars (Philadelphia: Univ. of Penn), 288

Chandrasekhar, S. 1933, MNRAS, 93, 462

Coates, D. W., Halprin, L., Sartori, P. A., & Thompson, K. 1983, MNRAS, 202, 427

Etzel, P. B. 1975, Master's Thesis, San Diego State Univ.

Etzel, P. B. 1980, EBOP User's Guide, 3rd. ed., Dept. of Astron. Univ. California, Los Angeles

Etzel, P. B. 1981, in Photometric and Spectroscopic Binary Systems, eds. E. B. Carling & Z. Kopal (Dordrecht: Reidel), 111

Giménez, A., & Díaz-Cordovés, J. 1992, Improving the Light Curve Synthesis Program EBOP: Variable Position of the Periastron and Second Order Limb Darkening, (these proceedings)

Lacy, C. H., Frueh, M. L., & Turner, A. E. 1987, AJ, 94, 1035

Lucy, L. B., & Sweeney, M. A. 1971, AJ, 76, 544

Nelson, B., & Davis, W. 1972, AJ, 174, 617

Popper, D. M. 1980, AR&AA, 18, 115

Popper, D. M. 1982, ApJ, 254, 203

Popper, D. M. 1984, AJ, 89, 132

Popper, D. M. 1986, PASP, 98, 1312

Popper, D. M. 1988, AJ, 95, 1242

Popper, D. M. 1990, AJ, 100, 247

Popper, D. M., & Etzel, P. B. 1981, AJ, 86, 102

Popper, D. M., Lacy, C. H., Frueh, M. L., & Turner, A. E. 1986, AJ, 91, 383

Reisenberger, M. P., & Guinan, E. F. 1989, AJ, 97, 216

Russell, H. N., & Merrill, J. E. 1952, Contr. Princeton Univ. Obs. No. 26

Thurman, G. M. 1990, Master's Thesis, San Diego State Univ.

Tomkin, J., & Popper, D. M. 1986, AJ, 91, 1428

Tümer, O., Ibanoglu, C., Tunca, Z., & Evren, S. 1984, Ap&SS, 104, 225.

Wilson, R. E. 1979, ApJ, 234, 1054

Wilson, R. E. 1990, ApJ, 356, 613

Wilson, R. E., & Devinney, E. J. 1971, ApJ, 166, 605

Discussion

Q (C. R. Chambliss): A comment about non-linear limb darkening coefficients. Even in the least complex and best observed systems, it is probably impossible to determine empirically non-linear limb darkening coefficients with existing models. It may be possible to use these coefficients as input parameters, but the reverse seems not to be viable.

A: Yes, even determining linear limb-darkening coefficients, requires very good observations and an eclipse geometry conducive to such an effort (see for example the discussion by Popper 1984). Some photometrists have been making advances in pushing the precision of ground based photometry down to the millimag level. If such observations were routinely possible, then empirical determinations could probably be made for non-linear coefficients. However, even now it should be possible, at least in a statistical sense, to test various non-linear models against the linear model (especially with multicolor, intermediate-band photometry) by noting systematic differences between them. In that way, we could at least justify the use of non-linear models.

Q (C. R. Chambliss): In the EBOP program, how well are non-complex classical Algol systems treated?

A: Not very well because of the perturbation scheme used to calculate the proximity effects. These perturbations for most Algols are too large for what is basically a spherical model. Still, the code can be used for exploratory solutions to examine data quality and provide starting elements for more rigorous models. For example, one of my graduate students, Yoshiko Narasaki, recently used EBOP on the system SS Ceti for which she had 1200 observations in each filter. She got reasonable values for i, r_p, k, and L_p. The distortions of the secondary star from a biaxial ellipsoid were so extreme, that a value of q= 1.6 was determined, which is inverted from what would be expected. (EBOP essentially swept its inadequacies under the rug by adjusting the value of q to fit the observations). However, the b-axis dimension of the secondary component implied a value of q=0.3, which was subsequently used as input to the Wilson-Devinney code for a very successful series of solutions.

Improving the Light Curve Synthesis Program EBOP: Variable Position of the Periastron and Second Order Limb Darkening

A. Giménez
Laboratorio de Astrofísica Espacial y Física Fundamental, Estación de Villafranca del Castillo, Apartado 50727, 28080 Madrid, Spain
J. Díaz-Cordovés
Instituto "Alvaro de Mendaña", 24400 Ponferrada (León), Spain

Abstract. EBOP is one of the most commonly used codes for the light curve synthesis of detached eclipsing binaries. We have added to the original version the possibility to analyze binaries showing fast apsidal motion by introducing a variable position of the periastron. On the other hand, we have also introduced the possibility of using non-linear limb darkening laws to reproduce the surface flux distribution. A systematic difference was found in the orbital inclinations computed with our extended version of EBOP with respect to the standard one.

The program EBOP (Eclipsing Binary Orbit Program) is based on the eclipsing binary model by Nelson-Davis-Etzel (NDE model) (Nelson & Davis 1972, Etzel 1980, Popper & Etzel 1981). We have adopted version 3, which includes modifications 1 to 15, as the standard code and two main improvements have been implemented.

We have considered the possibility that the position of the periastron in eccentric eclipsing binaries could vary during the observational campaigns. This has proven to be necessary to analyze the observations of certain binaries showing fast apsidal motion like, e.g., GL Car (Giménez & Clausen 1986), CW Cep (Clausen & Giménez 1991) or BW Aqr (Clausen 1991). The case of the eclipsing binary V477 Cyg was nevertheless the first analysis of a light curve in which we found ourselves forced to solve the problem of synthetic light curve codes with a variable position of the periastron (Giménez & Quintana 1992). The effect of apsidal motion on the light curve is a distortion and widening of the secondary eclipse. This can be seen in Figure 1, where we have plotted the secondary eclipse of the light curve of V477 Cyg by O'Connell (1970) using all his observations and a single linear ephemeris. An analysis of this data, using the standard program EBOP, will produce systematic effects on the final elements.

To take into account variable values of ω, we used Heliocentric Julian Dates as input, and introduced an equation of apsidal motion to calculate phases in subroutine OPHASE for each individual observation. An independent and accurate determination of the apsidal motion parameters should be available before the light curve analysis. Due to the increased number of significant digits that should be taken into account in all calculations, double precision arithmetics is required.

In order to check our extended version of EBOP, hereafter called EBOPC, we have studied the case of GL Car. In our previous analysis (Giménez & Clausen 1986) using EBOP and WINK (Wood 1972), we decided to make a separate analysis of the data collected in 1982 and 1983. The results of the analysis with EBOP are shown in Table I for the y filter and an adopted value of the ratio of radii, k = 0.95. With the same assumptions, but using the variable-ω facility with EBOPC, we obtained the results also included in Table I (last column) taking the whole data sample. Finally, using all observations but fixing ω to an average value, we obtained the results shown in column 4 of Table I by means of EBOP. These different fittings show that our adopted approach of splitting the observations gave good results, though all are superseded by the results with EBOPC. Unfortunately, in cases with observations over a larger span of time, the effect is more dramatic and, particularly, if we do not adopt a fixed k and try to solve for it as a free parameter.

On the other hand, available versions of EBOP use the classical linear limb darkening law,

$$I(\mu) = I(1) \{1 - u(1 - \mu)\} \qquad (1)$$

Several non-linear limb darkening laws have been proposed but the more generally adopted one is expressed in the form of a quadratic equation,

$$I(\mu) = I(1)\ \{1 - a(1 - \mu) - b(1 - \mu)^2\} \qquad (2)$$

Furthermore, we have recently introduced a new non-linear law (Díaz-Cordovés & Giménez 1992) with the form,

$$I(\mu) = I(1)\ \{1 - c(1 - \mu) - d(1 - \sqrt{\mu})\} \qquad (3)$$

This square-root law reproduces very accurately the theoretical values of the flux, as well as those of $I(\mu)/I(1)$, computed from the model atmospheres by Kurucz (1979) with radiative transport.

The NDE model uses the approximation by Binnendijk (1960, 1974), where the unnormalized light of each of the components is given, as a function of the orbital phase, by the equation,

$$\ell(\vartheta) = \ell(\text{quad})(1 - f(\varepsilon)\ \sin^2 i\ \cos^2\vartheta) \qquad (4)$$

being i the orbital inclination, ϑ the true phase angle measured from primary minimum, ε the stellar oblateness, and $\ell(\text{quad})$ the unnormalized light at quadrature. The function $f(\varepsilon)$ is computed, for $i = 90°$, as

$$f(\varepsilon) = 1 - \ell(\text{conj})/\ell(\text{quad}) \qquad (5)$$

We have computed $\ell(\text{conj})$ and $\ell(\text{quad})$, for oblate components, using the three limb darkening laws (1), (2), and (3) given above, and the following expressions were obtained:

$$\ell(\text{conj}) = \pi\ J\ b^2\ \{1 - u_1/3\ (1 + 4\varepsilon/5) - u_2/6\ (1 + 6\varepsilon/5) - u_3/5\ (1 + 8\varepsilon/9) + y\varepsilon\ (2 - 14u_1/15 - 8u_2/15 - 26u_3/45)\} \qquad (6)$$

$$\ell(\text{quad}) = \pi\ J\ ab\ \{1 - u_1/3\ (1 - 2\varepsilon/5) - u_2/6\ (1 - 3\varepsilon/5) - u_3/5\ (1 - 4\varepsilon/9) + y\varepsilon\ (3 - 13u_1/15 - 2u_2/5 - 23u_3/45)\} \qquad (7)$$

These equations are valid for the three limb darkening laws. Expressions for the linear approximation (1) are obtained taking $u_1 = u$, and $u_2 = u_3 = 0$. For the quadratic law (2), we should take $u_1 = a$, $u_2 = b$, and $u_3 = 0$. Finally, the square root law (3) requires that $u_1 = c$, $u_2 = 0$, and $u_3 = d$. The total light expressions, valid for linear and non-linear limb darkening laws, and the expression for $f(\varepsilon)$, given by equation (5), have been introduced in EBOP. Thus this new version, which is now available, allows the use of the non-linear limb darkening laws (2) and (3) to analyze the light curves of eclipsing binaries.

Popper & Etzel 1981 found that EBOP can be used to analyze detached systems with $<\varepsilon>$ around 0.04 or smaller. The standard versions of EBOP compute $f(\varepsilon)$ using Taylor expansions in ε of equation (5) and disregard terms with powers greater than 1. For the eclipsing system V1031 Orionis, whose ε values are around 0.01 and 0.03 for the primary and secondary components, the orbital inclination obtained with the standard version of EBOP is around $0^{\circ}.1$ larger than the value obtained using our new version which computes $f(\varepsilon)$ directly from expression (5). For systems with much less oblate components, like DM Virginis (around 0.002 for each component), the effect is undetectable. Special caution should be taken when EBOP is used to analyze eclipsing systems with values of ε close to the applicability limit given above, even if our extended version is used.

References

Binnendijk, L. 1960, "Properties of Double Stars", University of Pennsylvania Press. Philadelphia.
Binnendijk, L. 1974, Vistas in Astronomy, 16, 61
Clausen, J.V. 1991, A&A, 246, 397
Clausen, J.V., Giménez, A. 1991, A&A, 241, 98
Díaz-Cordovés, J., Giménez, A. 1992, A&A, in press
Etzel, P.B. 1980, "EBOP User's Guide", Department of Astronomy, University of California, Los Angeles.
Giménez, A., Clausen, J.V. 1986, A&A, 161, 275
Giménez, A., Quintana, J.M. 1992, A&A, in press
Kurucz, R.L. 1979, ApJS, 40, 1
Nelson, B., Davis, W.D. 1972, ApJ, 174, 617
O'Connell, D.J.K. 1970, Vistas in Astron., 12, 271
Popper, D.M., Etzel, P.B. 1981, AJ, 86, 102
Wood, D.B. 1972, "A Computer Program for Modeling Non-spherical Eclipsing Binary Systems", Goddard Space Flight Center, Report X-110-72-473, Greenbelt, Maryland

TABLE I
GL Car (y) EBOPC test runs (k = 0.95 fixed)

	EBOP			EBOPC
	1983	1982	both	both
J_B	0.960	0.962	0.954	0.959
	3	5	4	3
r_A	0.2190	0.2199	0.2178	0.2195
	5	6	5	4
i	86.37	86.48	86.44	86.43
	5	6	4	3
$e\cos\omega$	-0.1396	-0.1451	-0.1423	var.
$e\sin\omega$	-0.042	-0.013	-0.031	var.
$u_A=u_B$	0.30	0.30	0.30	0.30
y_A	0.37	0.37	0.37	0.37
y_B	0.38	0.38	0.38	0.38
σ	0.0057	0.0069	0.0098	0.0062

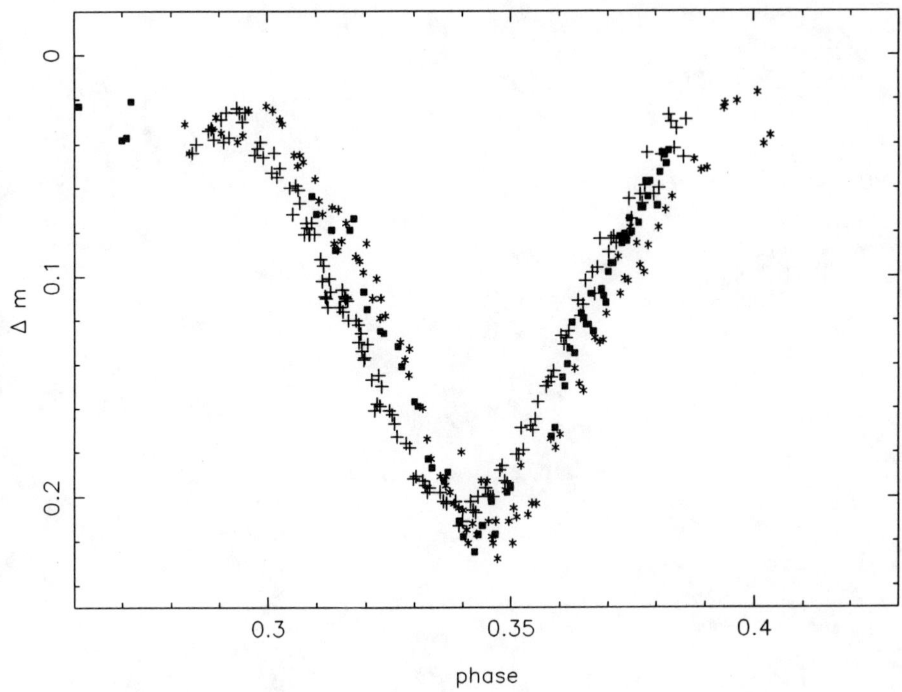

Figure 1

Secondary eclipse of V477 Cyg from observations by O'Connell (1970). Symbols denote measurements obtained in different years. A single linear ephemeris derived from primary eclipses was adopted.

Linear and Non-Linear Limb Darkening Coefficients for Synthetic Light Curve Modeling of Binary Stars

J. Díaz-Cordovés
Instituto "Alvaro de Mendaña", 24400 Ponferrada (León), Spain
A. Claret
Instituto de Astrofísica de Andalucía, Apartado 3004, 18080 Granada, Spain
A. Giménez
Laboratorio de Astrofísica Espacial y Física Fundamental, Estación de Villafranca del Castillo, Apartado 50727, 28080 Madrid, Spain

Abstract. Linear and quadratic limb darkening coefficients have been calculated by means of a least squares fitting for a large sample of stellar model atmospheres. A range in temperatures from 4000 to 50000 K has been covered from the zero-age main sequence up to the giant phases. A new law for the limb-darkening of hot stars, based on a square-root term, is also included. The validity of the different approximations for the analysis of light curves of eclipsing binaries is discussed.

The limb darkening is an important parameter in analyzing eclipsing binary systems. A linear limb darkening law usually has been adopted to analyze light curves of eclipsing systems. It is well known though that the limb darkening is not linear. A more general approximation is that given by a quadratic expression. We have introduced a new non-linear limb darkening law with a term of a power 1/2 of μ in the form (Díaz-Cordovés & Giménez 1992a),

$$I(\mu) = I(1)\,\{1 - c(1 - \mu) - d(1 - \sqrt{\mu})\} \qquad (1)$$

Monochromatic linear, quadratic, and square root coefficients have been calculated by means of a least-squares fitting using theoretical values of the emergent specific intensity from model atmospheres by Kurucz (1979) covering a range of temperatures from 5500 to 50000 K and surface gravities from 0.0 to 5.0 in cgs units. Details are given by Díaz-Cordovés & Giménez 1992b).

We have also computed linear and quadratic coefficients, by means of a least-squares fitting (Claret & Giménez 1990) using values of I(μ) calculated on the basis of the Uppsala Model Atmosphere (UMA) by Gustafsson et al. (1975). A range of effective temperatures from 4000 to 6730 K has thus been covered with log g values from approximately main-sequence stars up to the giants. We have integrated these limb darkening coefficients in the most commonly used photometric bands (uvby and UBV).

We have compared our results with those by Grygar et al. (1972), Al-Naimiy (1978), and Wade & Rucinski (1985). The differences between the linear coefficients given by several authors are produced, essentially, by the method used to fit the theoretical I(μ)/I(1) values rather than the adopted model atmospheres. The quadratic coefficients are less affected by the fitting method than the linear values.

Figure 1 compares, for a model atmosphere of 25000 K, the results obtained directly from the models by Kurucz (1979) with those derived from the approximations given by the three limb darkening laws mentioned above. Our empirical law (1) is the best choice for radiative model atmospheres (T_{eff} > 8500 K). This square-root law reproduces theoretical values of flux and specific intensity very accurately for any angle μ. The quadratic law is a better adoption for convective models, and both non-linear approximations are much better than the linear law. The least-squares fitting method does not impose that the approximate flux (F′) coincides with the theoretical flux (F) but the values of F′ computed using the two non-linear laws are almost indistinguishable from F. Even the linear law produces only small deviations in the values of the total flux.

We have made corrections in the EBOP program (Etzel 1980) so as to allow use of the non-linear limb darkening laws. We have used this version of EBOP to analyze synthetic observations of eclipsing binaries with spherical components and circular orbits. The synthetic observations were obtained using theoretical I(μ)/I(1) values computed from models by Kurucz (1979) and adding a realistic amount of ramdom errors. Data were analyzed using the three limb darkening approximations and two types of coefficients for the linear and quadratic laws: those published by Wade & Rucinski (1985) and our coefficients computed by least squares.

Table 1 shows the results from the analysis of a synthetic light curve with "observation errors". The values of J_s, r_p and k are little affected by the adopted limb darkening law or the fitting method used to compute the coefficients. The orbital inclination is the most affected parameter, and the larger deviation is obtained when Lin.WR is used. The solution using the square root law reproduces the theoretical input values very accurately but, unfortunately, typical observation errors prevent the detection of the systematic effects that the different limb darkening laws produce in the residuals of the solutions of the light curves. Test runs made with theoretical light curves without observational "errors" clearly show that in most cases the best results are obtained with the square root non-linear law or, for a linear fit, with the least-squares values. A more detailed analysis is given by Díaz-Cordovés & Giménez (1992b).

There is some controversy about which is the most adequate fitting method to compute linear limb darkening coefficients to analyze eclipsing systems. We think that the controversy is about an impossible problem: to find the optimal linear approximation to a non-linear effect. No chosen fitting method will be optimal. The correct solution is to adopt an adequate non-linear limb darkening approximation. Our new empirical limb darkening law (1) is found to be the most adequate for radiative atmospheres.

References

Al-Naimiy, H.M. 1978, Ap&SS, 53, 181

Claret, A., Giménez, A. 1990, A&A, 230, 412

Díaz-Cordovés, J., Giménez, A. 1992a, A&A, in press

Díaz-Cordovés, J., Giménez, A. 1992b, in preparation

Etzel, P.B. 1980, "EBOP User's Guide", Department of Astronomy, University of California, Los Angeles

Grygar, J., Cooper, M.L., Jurkevich, I. 1972, Bull. Astron. Inst. Czech., 23, No 3, 147

Gustafsson, B., Bell, R.A., Eriksson, K., Nordlund, A. 1975, A&A, 42, 407

Kurucz, R.L. 1979, ApJS, 40, 1

Wade, R.A., Rucinski, S.M. 1985, A&AS, 60, 471

Table 1.

	Theoretical	Lin.LS	Lin.WR	Quad.LS	Quad.WR	Sq.Root
J_s	0.8960	0.8953	0.8953	0.8955	0.8944	0.8956
		20	19	19	19	20
r_p	0.119	0.1187	0.1191	0.1188	0.1188	0.1189
		5	5	5	5	5
k	0.91	0.914	0.905	0.913	0.914	0.911
		8	7	7	7	7
i	89	88.96	89.13	88.97	88.95	89.01
		4	5	4	4	4
SFACT	1.0	0.9997	0.9996	0.9997	0.9997	0.9997
σ		0.0049	0.0049	0.0049	0.0048	0.0049

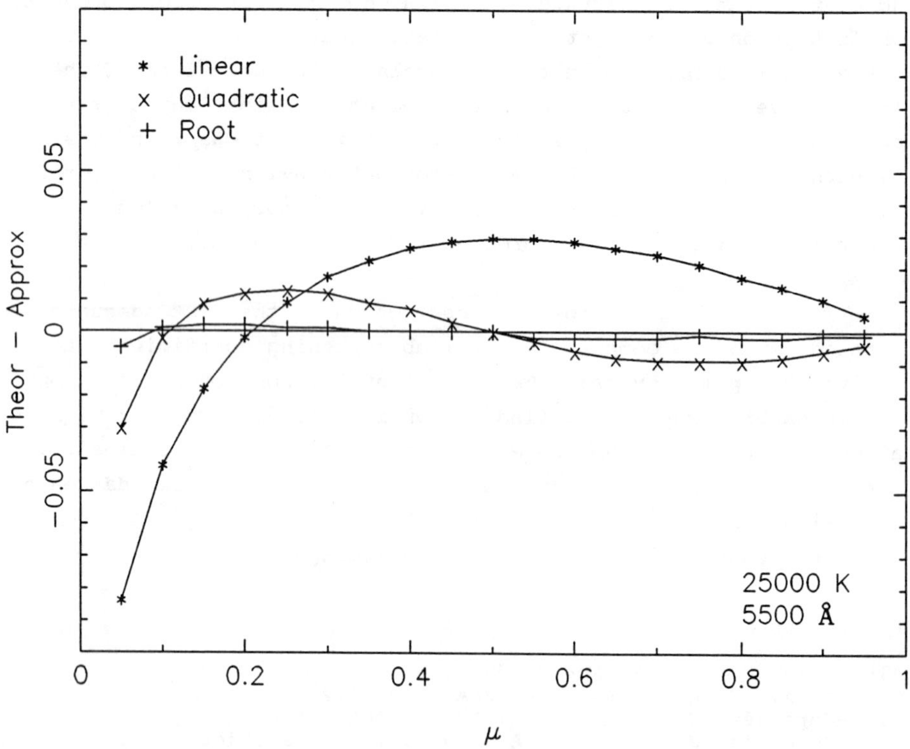

Figure 1.

Differences between Theoretical and Approximate $I(\mu)/I(1)$ values for a model atmosphere by Kurucz (1979) with log g = 4.0 and solar chemical composition.

Light2: A Light-Curve Modeling Program

Graham Hill
Dominion Astrophysical Observatory, Herzberg Institute of Astronomy, National Research Council, 5071 W. Saanich Road, Victoria, British Columbia V8X 4M6, Canada
Slavek Rucinski
Space Astrophysics Laboratory, Institute of Space and Terrestrial Science, Department of Physics, Petrie Building 206, York University, 4700 Keele Street, Toronto, Ontario M3J 1P3, Canada

Introduction

Since 1968 when the first descriptions of computer codes capable of modeling eclipsing binary light curves appeared in the literature (Lucy 1968, Hill and Hutchings 1970, Wilson and Devinney 1971, Wood 1971, 1972 and Rucinski 1973), many developments have taken place. Other programs have been written over the years and this volume will provide descriptions of many of them. These codes are now capable of generating more than light curves, *e.g.* the Wilson and Devinney code (WD hereafter) has been modified to calculate the effects of tidal distortion and the reflection effect on binary-star velocity-curves (Wilson and Sofia 1976). Broadband fluxes based partly on observation and partly on theoretical colours are incorporated into the program LIGHT2 (Hill 1979, Rucinski 1973, Hill 1985). The use of computer codes to aid in the measurement of spectra was first broached by Van Hamme and Wilson (1985) and a variant of it has been implemented by Hill (see Hill, Fisher and Holmgren 1989b). Spot models are incorporated into many codes (*e.g.* Eaton and Hall 1979, Budding and Zeilik 1987, Kang and Wilson 1989 and Hill, Fisher and Holmgren 1990). Limb darkening values, originally limited to linear approximations, have been replaced by interpolation formulae based on model atmosphere results, *e.g.* Hill (1979) analyzed the Carbon and Gingerich (1969) tables; results which were later supplanted by the work of Wade and Rucinski (1985) who created similar interpolation coefficients based on the atmospheres of Kurucz (1979). In this present volume Wilson expands on ideas to include the calculation of polarization parameters, absorption line indices, radial and rotational velocities. and so on. In addition to the increased realism of the computer codes, researchers have also moved towards a more automatic analysis of light curves as well as a shift in observing priorities. In this regard we note that the trial and error methods of solution have given way to more or less automatic fitting once a reliable first guess has been made. These fitting techniques were originally based on Newton's method (Wilson 1979) or the method of Marquardt (Hill 1979), and recently the simplex algorithm (Kallrath and Linnell 1987) has been used. Linnell (1989) has also used a differential correction routine with a modification of the derived correction factors to provide better convergence. In all cases, once a solution is found the question of uniqueness always remains, since these methods depend on finding a minimum chi-squared in the hyperspace that has many local minima. Proctor and Linnell (1972) were the first to use simulated light curves to test the reliability of a solution as did Hill (1979). In his present code Linnell(1989) continues with this approach.

In part, the question of uniqueness hinges on the available mass ratio since this may be only weakly determined from a photometric solution, particularly for detached systems or others that only undergo partial eclipses. We feel that a very necessary part of any light curve analysis is the related spectroscopic study for without it a light-curve solution often needs to be redone once a reliable mass ratio is available. Fortunately there is a developing trend in which a number of photometric observers are moving into the spectroscopic realm to garner the spectroscopy needed for a complete analysis of existing light curves. Because the spectroscopy and photometry are being combined, progress is being made to improve the measurement of radial velocities (RV) to match the increased realism of available light-curve computer-codes. These developments bode well for the future of this field which, while it has matured (like many of us!), is still showing the vitality needed for future advances.

In this paper we will discuss the evolution of LIGHT2 and its limitations. In addition we indicate areas that we deem important for future improvement. The code is available to potential users and versions are currently running in VAX and SUN environments.

The Model

a) LIGHT and WUMA3

LIGHT2 is made up of two program, LIGHT (Hill 1979) and WUMA3 (Rucinski 1973, 1976). We discuss each in turn.

LIGHT: The basis of this program is the Roche model combined with Wood's (1971, 1972) elegant Gauss-Legendre quadrature scheme. A similar approach was used by Rucinski (1969) who, in studying proximity effects in close binaries, coupled Roche geometry to a Gauss-Legendre-Chebeyshev quadrature. Initially LIGHT was developed using this scheme but it was abandoned in favour of the Gauss-Legendre method because of a failure of the Chebyshev quadrature (used in one coordinate) to improve the integration accuracy as the order was increased. As used by Wood on a triaxial ellipsoidal model it was very fast and very accurate even when using only 100 points over the surface of a star. When applied to Roche geometry the quadrature is not so straightforward, since it is based on knowing the integration limits on each star. For Wood's model these are analytic functions but for Roche geometry they must be calculated by algorithms which can be awkward and complicated. This is a real limitation of the scheme which is still very fast and accurate when only 256 points are used over the surface of each distorted star but which becomes very difficult when over-contact systems or spotted stars are to be modeled. For this reason WUMA3, the code of Rucinski (1973, 1976), was integrated within LIGHT to create the new program LIGHT2.

WUMA3: This integration scheme is based on the model of Lucy (1968) and requires many grid or integration points (~2500) to achieve an integration accuracy that will match the observations (rms ~ .005 mag). Originally the use of this type of integration meant long runs to yield a solution but with the availability of VAXes, and more recently SUN's, execution time is not an issue. LIGHT2 runs in both of these environments. The original quadrature devised by Lucy was specifically applied to contact and over-contact systems so to incorporate WUMA3 into LIGHT2 the potentials were generalized to allow detached and semi-detached configurations to be treated. As a result, within LIGHT2, the use of LIGHT or WUMA3 is transparent to the user inasmuch as they share the same physics and light-curve solver (CURFIT, Bevington 1969). As with the development of LIGHT we have made extensive checks between these two programs and the results are good to <0.001 mag (see Hill 1979 also). In the course of these checks a number of "bugs" in LIGHT were revealed though none would have affected previous results. In addition, both codes have been extensively checked against the WD code by the group in St Andrews Scotland (Bell and Hilditch, private communication and Bell *et al.* 1991) and all three codes are in good accord. This is a satisfying situation considering the somewhat differing physical assumptions between WD and LIGHT2, particularly in the treatment of the reflection effect. The bonding of WUMA3 with LIGHT has been very useful since it is very easy to add spots to WUMA3 and the quadrature is readily expandable to incorporate extended atmospheres, gas streams and additional disks though, following current practice, we will not implement them until we have a direct need.

b) Atmospheres

The atmospheres used are either black body intensities or a hybrid scheme based upon black body intensities, observed (UBVRI, uvby) colours and theoretical fluxes. In this latter scheme, the run of (U-B), (B-V), (V-R), (R-I), (u-b), (v-b), (b-y) with spectral type has been derived from a number of sources in the literature and the spectral type dependence is then converted to that of T_e by the use of Popper's (1980) calibration. The relation between passband (U,B,...) and T_e results by applying the calibration of V and v with T_e for log g=4 as defined by Buser and Kurucz (1978) and Relyea and Kurucz (1978). These fluxes are then converted to normal emergent intensity by normalizing them by a factor dependent on the limb darkening $u(\lambda)$. Here, for linear limb darkening, Passband(intensity) = Passband(flux) x $(1-u/3)/(2 \times \pi)$. Thus, given a temperature and λ, the black body intensity is calculated and then multiplied by the appropriate passband intensity at that temperature. This procedure is described in detail by Hill (1979, 1985). Obviously this whole approach can be modified to use the fluxes predicted by model atmospheres, provided they can be converted to intensities by applying suitable limb darkening factors, but the hybrid scheme does have the advantage that it is at least partially based on observation. Ultimately this scheme will be replaced by one using Kurucz's (1992, present volume) latest atmospheres and limb darkening grids as a function of μ, λ, T_e, log g and abundance.

c) Limb darkening

The values of limb darkening as a function of λ, T_e and log g used in LIGHT2 are based on an analysis of the tables of Carbon and Gingerich (1969). This analysis was both linear and quadratic in scope and is described by Hill (1979). In time LIGHT2 will incorporate the second order (quadratic interpolation coefficients) results of Wade and Rucinski (1985) which are based on the LTE atmospheres of Kurucz (1979). Vaz and Nordlund (1985), in their discussion of the reflection albedo (see next section), have also generalized the effects of the reflection effect on limb darkening and have also implemented these results in Wood's code (WINK). Their whole analysis is based on the study of cooler components but it would be of interest to emulate their work and include it in LIGHT2. Peraiah and Rao (1983) have also made a start in investigating the effects of the reflection effect on limb darkening and it is to be hoped that this work also would be continued.

d) Reflection effect

This effect, that is the run of incoming flux (S) and incident angle (μ_0) over the surface of the heated star, is calculated using an integration scheme of rings and sectors after Hutchings (1968), and Hill and Hutchings (1970). Rather than try to evaluate (S, μ_0) at each integration point over the heated surface of each star, it is instead calculated at a series of points along planes running from the substellar point through the pole and also around the equator. The value of (S, μ_0) at any given point is then given by interpolation. To speed things up an interpolation formula was fitted to a series of results for detached systems such that they reproduce the "exact" values to within 5%. Then to convert the incoming flux to a temperature, use is made of Sobieski's (1965) formulation. Here the source function due to the incoming light, itself a function of the bolometric heating and the incident angle of the incoming light, is added to that of the heated star, although the way that an atmosphere reacts to external heating still requires study. To this end a beginning was made by Buerger (1972) who simply modified the external boundary condition to simulate incoming radiation in the program ATLAS (Kurucz 1969) and then generated a new model atmosphere. Later Vaz and Nordlund (1985) re-examined this problem and were able to generalize some of their results. They found that the albedo was not single-valued but varied itself with incident angle. In principle both Buerger's approach and that of Vaz and Nordlund could work in our codes but such rigorous treatments would demand the calculation of many thousands of atmospheres for each phase configuration. In addition, the heating effect must surely generate mass motions or circulation currents, the scale and observational consequences of which we don't as yet know. The current situation is nicely summarized in a comprehensive paper by Wilson (1990).

e) Eccentric orbits

In treating these orbits we scale the relative radii at each phase point by the inverse of the separation and henceforth that phase is treated as if the eccentricity were zero, with the mean anomaly (photometric phase) being replaced by the true anomaly. This implies that the star adjusts instantaneously to separation changes, and hence shape changes, and that the polar radius is constant. Wilson (1979) assumes that the volume is constant, an assumption which is valid to first order (Avni 1976). In eccentric orbits the heating will change with separation, so one generally chooses the interpolation formula to calculate the reflection effect anew at each phase since calculations involving the system of rings and sectors can be slow.

f) The light-curve solver

As with the WD code, Lucy (1973) and Linnell (1989), the light curves are solved by differential corrections. LIGHT2 uses the program CURFIT (Bevington 1969) based on the method of Marquardt (1963). This program is described in great detail by Bevington and is a excellent tool for solving light curves. However, because of the need to calculate derivatives, it may not be the optimum tool and we refer the reader to the simplex method as applied to light curves (Kallrath and Linnell 1987) or other algorithms described for example by Luenberger (1973) and Wolfe(1978).

The most difficult aspect of a differential correcting program is the calculation of derivatives, since they are time consuming and subject to round-off errors. Within LIGHT2 the derivatives are generally calculated using a lower order quadrature (usually a factor of 2) than in the light curve generator itself, and the derivatives are then obtained by differencing these less accurate light curves. In doing this we are following Lucy's (1973) precept of using the lowest accuracy that one can get away with. His comment, however, was made at a time when, by today's standards, computers were very limited both in core memory and their execution speeds, factors which are not that important due to the advent of SUN's and VAXes. Thus this restriction on accuracy is largely self-imposed and could be a factor in determining how unique a particular solution is. We suggest that after a solution is obtained that a subsequent one be run with derivatives as accurate as possible so as to check for the sensitivity of a solution to this effect. The question of uniqueness is a difficult one in that one must find a global minimum in the chi-squared, parameter hyper-plane which is replete with local minima. Rafert and Markworth (1986), using the WD code, outline a procedure that will help in reaching a global minimum. Alternatively, Kallrath and Linnell (1987) have applied the simplex algorithm to the solution of light curves to avoid this problem. When using a differential corrector one should use a series of differing starting values in order to see what other solutions, or apparent, solutions are present. Because some parameters only weakly affect a light curve, a certain amount of caution must be used in solving for them. Limb darkening and the system mass ratio immediately spring to mind. Where possible these respective quantities should be fixed to either the theoretical values or the observed one. With regard to the mass ratio, in our experience it can be very poorly determined from a light-curve solution but in the absence of an observed value including it as an unknown is justified. This attitude is strongly challenged by a referee and we quote from his report. "Hill (plus Popper and some others) have made the assertion several times that the only worthwhile mass ratio is a radial velocity mass ratio. This is just incorrect. Granted, the photometric mass ratio is only weakly or not at all determined for detached binaries, but it can be very strongly determined for semi-detached and over-contact cases, especially with complete eclipses...... The actual situation is that sometimes the photometric mass ratio is much better than the velocity mass ratio". In response to this one of us (G.H.) feels that the reaction to the velocity mass ratio is a product of the past when spectroscopic measures were suspect. Currently, through the use of the cross-correlation method, velocity mass ratios can be quite reliably determined for all systems with favourable light ratios. What we see in the literature are attempts to determine photometric mass ratios for patently unsatisfactory situations involving for example partial eclipses and/or detached systems; a situation that elicits a strong negative response from spectroscopists.

Most of the current codes allow for a simultaneous multicolour solution. Here the average of the colour-dependent derivatives is used to correct the elements. If a model poorly reproduces a particular passband, and generally the U passband of the UBV system creates problems, then a solution may be skewed by these data. Thus occasionally one will arbitrarily omit a passband from a solution. Difficulties with a passband generally indicate a problem with the adopted physics, notably in the treatment of the reflection effect where the wavelength dependence of the redistributed light may be quite different from the simple model adopted here (see Buerger 1972).

Wilson (1979) combines the light and velocity solutions in a generalized solution. Such a method could be incorporated within LIGHT2 but this would require the use of a set of velocity measures that themselves depend on the model. We will return to this in h).

g) Spot models

We incorporated spots into WUMA3 to facilitate an analysis of ER Vul (Hill, Fisher and Holmgren 1990). Up to ten may be added to each star though, if needed, a differing distribution could also be implemented. The location of the spots, their sizes, shapes (circular or elliptical), and the temperature difference relative to the local photosphere can all be specified. The spots are assumed to have no penumbra. When spots are added the resultant light curve can be of reduced accuracy if the added features are small and hence are defined by only a few grid points. However, to increase the accuracy of the integration would require a large increase in the total number of grid points. This effect could be severe in a deep eclipse when a spot is eclipsed but this can only be checked by trial and error for a given situation. When good data are available it would be worthwhile to add time dependent positioning of the spots, to investigate such things as drifts to higher latitudes caused by differential rotation or drifts in longitude resulting from non-synchronous rotation. The location and temperature difference for each spot may also be included as unknowns in the general solution.

h) *Radial velocity corrections*

The calculation of velocity corrections is needed to render the observed RV's Keplerian, since the measures are affected by the asymmetric effects of eclipses, distortion and heating. These effects were calculated by Hill and Hutchings (1973) in their study of MR Cyg and are simply generated by associating a velocity with each grid point and then calculating a flux-weighted average over the surface of each star. Quite independently Wilson and Sofia (1976) and Wilson (1979) have incorporated these calculations within the WD code and included velocity measures along with the light curve to allow a generalized solution. While the velocity corrections improve the mass ratio, they do not help the measurement of the velocity that ultimately gives rise to the mass ratio. Specifically this method does not aid one in measuring the asymmetric profiles resulting from these proximity effects. Hutchings, in the paper by Hill *et al.* (1971), discusses this question, which really had no resolution until a suggestion by Van Hamme and Wilson (1985) created what we see as a breakthrough in the measuring procedure. This suggestion, which involved the generation of a synthetic spectrum from a light curve code and its use as a mask to measure RV's in the cross-correlation process, spurred the work of Hill, Fisher and Holmgren (1989b) which we will now briefly describe. In this case, rather than use a synthetic spectrum as a mask or a cross-correlation template-star it would be better to generate line profiles for each component at each phase point and use these profiles to measure a cross-correlation function (ccf) generated normally with reference to a reliable standard spectrum. Thus, once a ccf is available, one fits these profiles to the display and the velocities which result, though model dependent, are true Keplerian velocities. Hill *et al.* (1989b) also show how well these profiles match observed ccf's. The reader may think that this is a digression, discussing the measurement of RV's in a paper such as this, but the fact remains that a really significant advance in the measurement of RV arose out of a discussion of light curves by Van Hamme and Wilson. In our minds the measurement of RV's, particularly in W Uma systems, is completely linked to the solution of their light curves. For this reason we cannot use the light and velocity data to perform a simultaneous solution in the manner given by Wilson (1979). Fortunately the problem can easily be solved by successive iterations, i.e. we simply measure the velocities by fitting Gaussian profiles to the ccf's to derive a mass ratio. Solve the light curve with that mass ratio. Use the solution to generate a set of line profiles to remeasure the ccf's and solve the light curves anew. After one such iteration the process converges. This method is presently only used to analyze contact systems where the ccf's for the secondaries are very well defined. In semi-detached systems the secondary's ccf are not as well defined and they are usually measured by fitting Gaussian profiles to the peaks. These stars could readily be solved by Wilson's simultaneous analysis. In presenting their results, spectroscopists who use these measuring techniques should publish both the Gaussian velocities and those that depend on a model. We turn now to the calculation of the line profiles.

i) Line profiles

In LIGHT2 the line profiles are calculated under the following assumptions:

1) For each grid point Gaussian profiles of full-width half-maximum (FWHM) approximating the spectrograph resolution were shifted in velocity to the projected rotational velocity of the star (see Kopal 1959, Kopal and Kitamura 1968) and summed to create an intensity-weighted profile calculated for the visible surface of each star.

2) The limb darkening in the line may be varied, though in the absence of anything else we use the photospheric limb darkening value. Note that this may not be a very good assumption, since the line profiles for B stars calculated by Stoeckley and Mihalas (1973) show changes in shape and strength as a function of emergent angle that are hard to predict. In fact within this current volume, Kurucz shows that the limb darkening in the line is not the same as in the photosphere.

3) The effects of heating were included. [Note that in the paper of Hill *et al.* (1989b), the line-profile calculation was limited to contact or under-contact systems but with the incorporation of WUMA3 within LIGHT2 this is no longer a restriction.] Surprisingly, the line profiles calculated under these assumptions provide remarkably good fits to observed ccf's, but they are not adequate for the analysis of OB stars where Stark broadening is present. A useful development would be the incorporation of a grid of intensity profiles within LIGHT2. Such a database would need to be quite comprehensive, both in the lines used (He, C, Mg, O and Si), and in the available range of log g, T_e and abundance. Suitable ranges would be 2.0 to 4.5 in log g, 9000K to 50000K in T_e and for abundances of solar and 0.1 solar composition. Note that in the LMC and SMC, metal abundances are lower than solar and hence such models would have a direct relevance to current research, particularly that of eclipsing binaries in the Clouds (see Bell *et al.* 1991). Given such a database one could also generate rotational profiles for single stars that could be matched against high signal-to-noise reticon observations of rotational velocity standards and which would enable the pioneering work of Slettebak *et al.* (1975) to be re-examined and improved upon. Note that velocities derived from their work are only reliable to +/-20 km/s, since that is the quoted accuracy of their calibration, but with the use of high signal-to-noise reticon profiles coupled with more modern atmospheres and modelling procedures one should be able to improve the calibration a great deal.

j) Presence of a tertiary

When a third body is present, *e.g.* in the light curve of Algol, 44 i Boo or VW Cep, its light contribution can be handled in the light curve calculation in either of two ways; 1) a relative radius coupled with a temperature may be specified, or 2) the data themselves can be corrected for a known magnitude difference between the tertiary and the binary. This latter approach was used by Hill *et al.* (1989a) in their study of 44 i Boo., whereas Hill and Hutchings (1970), in their investigation of Algol, used a combination of the two methods . In this latter case they used an effective temperature and magnitude difference to deduce the relative radius of the tertiary.

Further Comments

a) Operation of LIGHT2

Some operations of LIGHT2 such as the setting of parameters, switching options on or off, and the selection of unknowns to be derived are controlled by the use of keywords and associated numbers or file names. This system, which is termed the COMMAND SYSTEM (Hill 1986), is used throughout the REDUCE suite of software (REDUCE, Hill *et al.* 1982; VCROSS, Hill 1982, 1986, 1990a) and avoids the complicated and generally unintelligible format statements used in the FORTRAN language. While the command system is great for control, it is not very useful for handling data because of the sheer volume one encounters. Data are therefore read with unformatted statements, though the data stream may be interrupted by keywords. For example, if one switched from phases and magnitudes to Julian dates (JD) and magnitudes these keywords would then give the necessary information to convert the subsequent JD's to phases.

b) Use of LIGHT2 at the DAO

LIGHT2 is generally used as a tool for the spectroscopy program that has been underway at the DAO since 1985. Over this time spectroscopy of more than 70 systems brighter than 8.5 mag has been completed using the 1872 element reticon on the 1.2m telescope, which yields spectra of resolution about 1 Angstrom (0.3Å/pixel) and a wavelength coverage of 500Å (4000 to 4500Å). These spectra are measured by cross-correlation to produce reliable orbits and mass ratios. The techniques of measuring the resultant ccf's by line profiles produced by LIGHT2 came from light curve studies. For late-type systems, given a preliminary mass ratio, a light curve solution is made from which one generates line profiles that in turn are used to measure the RV's anew. The revised mass ratio then leads to an improved light curve solution. This iteration then closes very quickly. For early-type systems LIGHT2 plays no such role but as noted above, once access is available to a database of intensity profiles for a large number of log g, T_e and abundance values we can use LIGHT2 to aid in the measurement of RV. Once the orbits are available they lead to the final physical parameters and their errors generated by the auxiliary program PHYSERR.

c) Auxiliary programs

To aid in the analysis of light curves for which there are many (>600) points, we condense the data to normal points using the program NORMAL (Hill 1990b). In essence this program fits a cubic spline to the data and then the normals are evaluated at the desired phases. This type of program was first developed by Hilditch (private communication), who also created a related program for the reduction of photometric observations where the skies were interpolated using splines and then later the comparison star observations were similarly interpolated at those times corresponding to the variable observation, thus allowing the magnitude differences to be found. Light curve solutions produced by LIGHT2 are plotted by LIGHT2P (Hill 1985) and from these plots come the diagrams that are suitable for publication.

A print-plot option is also included within LIGHT2 so that systems may be projected onto paper. This option is invaluable when spots are being added since it always seems difficult to interpret the coordinate systems that govern their placement (even when you write the software yourself!).

d) Some further problems

A problem that bedevils the analyses of light curves involves the value of the gravity brightening parameter (β), in the transition realm near spectral type F2 where the atmospheres change from being radiative to convective as one shifts to later spectral types. The clarification of this issue is slowly unfolding and it appears that the convective value of β should lie between 0 and 0.08 (*e.g.* Anderson and Shu 1977, Eaton *et al.* 1980 and Rucinski 1989). Note that the term gravity brightening is used rather that gravity darkening since, when local gravity increases, the surface brightness increases (see Anderson and Shu and also Rucinski 1989).

In addition, the actual value of the albedo, that is the efficiency of the atmosphere in absorbing and re-radiating incident light, is often included as a parameter. The usual value is 50% (see Rucinski 1969, 1989) but it is often changed to fit observations; *e.g.* see the discussion of TV Cas by Khalesseh and Hill (1992). The theoretical basis for calculating the enhanced temperature induced by the reflection effect must still be studied even though an excellent start has been made by Vaz and Nordlund. Perhaps it is unrealistic to expect that results could be generalized to allow for the accurate calculation of the emergent intensity without the calculation of an atmosphere at each externally illuminated grid point.

Summary

LIGHT2 is currently able to solve various types of eclipsing binaries ranging from detached to over-contact systems. It can also generate line profiles that have become invaluable tools for the measurement of RV's by cross-correlation methods. We expect to upgrade the code to incorporate the new limb-darkening coefficients by Wade and Rucinski. In addition, this change will see improvements to the modified black-body intensities that we can currently use. The inclusion of line profiles for early-type systems would be a worthwhile addition since it would aid in their RV measurement. Given such a database the code could be expanded to enable the rotational velocity calibration of Slettebak *et al.* for single stars to be refined, particularly when we have at hand many spectra of rotational velocity standards. We also contemplate writing a version that would combine the simplex algorithm with the method of Marquardt so that the uniqueness of a given solution could be explored. Since LIGHT2 is relatively easy to develop, potential users with interests not addressed here may want to contact us. Finally, as noted above, the code is available to potential users and versions are currently running in VAX and SUN environments.

References

Anderson, L. and Shu, F.H. 1977, *ApJ* **214**, 798.

Avni, Y. 1976, *ApJ* **209,** 574.

Bell, S.A., Hill, G., Hilditch, R.W., Clausen, J.V., Reynolds, A.P. and Gimenez, A. 1991, *MNRAS* **250,** 119.

Bevington, P.R. 1969, *Data Reduction and Error Analysis for the Physical Sciences* (McGraw-Hill: New York).

Budding, E., and Zeilik, M. 1987, *ApJ* **319,** 827.

Buerger, P. 1972, *ApJ* **177,**657.

Buser, R. and Kurucz, R.L. 1978, *A&A* **70,** 555.

Carbon, D.F. and Gingerich, O.J. 1969, *Theory and Observation of Normal Stellar Atmospheres,* ed. O.J. Gingerich (M.I.T. Press: Cambridge).

Eaton, J.A. and Hall, D.S. 1979, *ApJ* **227,** 907.

Eaton, J., Wu, C.-C. and Rucinski, S.M. 1980, *ApJ* **239,** 919.

Hill, G. 1979, *PDAO* **15,** 297.

Hill, G. 1982, *PDAO* **16,** 59.

Hill, G. 1985, *LIGHT2 User Manual.*

Hill, G. 1986, *COMMAND SYSTEM User Manual.*

Hill, G. 1990a, *VCROSS User Manual.*

Hill, G. 1990b, *NORMAL User Manual.*

Hill, G., Barnes, J.V., Hutchings, J.B., and Pearce, J.A. 1971, *ApJ* **168**, 443.

Hill, G., and Fisher, W.A. 1986, *PDAO* 16, 159.

Hill, G., Fisher, W.A., and Holmgren, D. 1989a, *A&A* **211,** 81.

Hill, G., Fisher, W.A., and Holmgren, D. 1989b, *A&A* **218,** 152.

Hill, G., Fisher, W.A., and Holmgren, D. 1990, *A&A* **238,** 145.

Hill, G., Fisher, W.A., and Poeckert, R. 1982, *PDAO* **16,** 43.

Hill, G., and Hutchings, J.B. 1970, *ApJ* **162,** 165.

Hill, G., and Hutchings, J.B. 1973, *A&A* **23,** 357.

Hutchings, J.B. 1968, *MNRAS* **141,** 329.

Kalesseh, B. and Hill, G. 1992, *A&A* (in press).

Kallrath, J. and Linnell, A.P. 1987, *ApJ* **313,** 346.

Kang, Y.W. and Wilson, R.E. 1989, *AJ* **97,** 848.

Kopal, Z. 1959, *Close Binary Systems* (John Wiley and Sons: New York).

Kopal, Z. and Kitamura, M. 1968, *Advances in Astronomy and Astrophysics* **6**, 125.

Kurucz, R.L. 1969, *Theory and Observation of Normal Stellar Atmospheres* ed. O.J. Gingerich (M.I.T.Press: Cambridge).

Kurucz, R.L. 1979, *ApJS* **40,** 1.

Linnell, A.P. 1989, *ApJ* **342,** 449.

Lucy, L.B. 1968, *ApJ* **151,**1123.

Lucy, L.B. 1973, *ASpSc* **22,** 381.

Luenberger, D.G. 1973, Introduction to Linear and Nonlinear Programming (Addison-Wesley Publ: Mass).

Marquardt, D.W. 1963, *J. Soc Ind. Appl. Math.* **11,** 431.

Peraiah, A. and Rao, M.S. 1983, *JApA* **4,** 175

Popper, D.M. 1980, *ARAA* **18,115.**

Proctor, D.D. and Linnell, A.P. 1972, *ApJS* **24,** 449.

Rafert, J.B. and Markworth, N.L. 1986, *AJ* **92,** 678.

Relyea, L.J. and Kurucz, R.L. 1978, *ApJS* **37,** 45.

Rucinski, S.M. 1969, *AA* **19,** 125.

Rucinski, S.M. 1973, *AA* **23,** 79.

Rucinski, S.M. 1976, *PASP* **88,** 777.

Rucinski, S.M. 1989, *Comments on Ap* **14,** 79

Slettebak, A., Collins, G.W., Boyce, P.B., White, N.M., and Parkinson, T.D. 1975, *ApJS* **29,** 137

Stoeckley, T.R. and Mihalas, D. 1973, *NCAR Technical Note NCAR-TN/STR-84.*

Sobieski, S. 1965, *ApJS* **12,** 263.

Van Hamme, W. and Wilson, R.E. 1985, *A&A* **152,** 25.

Vaz, I.P.R. and Nordlund, A. 1985, *A&A* **147,** 281.

Wade, R.A. and Rucinski, S.M. 1985, *A&AS* **60,** 471.

Wilson, R.E. 1979, *ApJ* **234,** 1054.

Wilson, R.E. 1990, *ApJ* **356,** 613.

Wilson, R.E. and Devinney, E.J. 1971, *ApJ* **166,** 605.

Wilson, R.E. and Sofia, S. 1976, *ApJ* **203,** 182.

Wolfe, M.A. 1978, *Numerical Methods for Unconstrained Optimization* (Van Nostrand Reinhold Co.: New York).

Wood, D.B. 1971, *AJ* **76,** 701.

Wood, D.B. 1972, *A Computer Program for Modeling Non-spherical Eclipsing Star Systems,* Goddard Space Flight Centre, Greenbelt, Maryland.

Binary Maker 2.0 - An Interactive Graphical Tool for Preliminary Light Curve Analysis
"Double Stars for the Masses"

D.H. Bradstreet
Department of Physical Science, Eastern College, 10 Fairview Drive, St. Davids, PA 19087-3696, USA

Abstract

An extensively revised version of the ***Binary Maker*** code has been written in the C language for MSDOS machines with VGA monitors which allows users to construct 3D Roche geometry visualizations of nearly any circular orbit close binary. The program revolves the system and simultaneously creates accurate light and radial velocity curves and plots them on the screen as the user watches the stars revolve. There is also a mode which allows synthetic line profiles to be generated. The user interactive nature of this program greatly facilitates preliminary light curve modeling, especially when modeling starspot regions, and the program also has tremendous potential as a teaching tool in light curve synthesis. The program is described in some detail and several examples are given to illustrate its use.

Introduction

A common thread in light curve analysis is the need for a relatively closely fitting preliminary model to the data which will serve as the starting point for a program which will then iterate towards a better solution (e.g., the differential corrections mode (DC) of the Wilson-Devinney program (Wilson and Devinney (1971); Wilson (1979) and later revisions). It can be quite tedious and time consuming to develop this first model, especially if the system is complicated by asymmetries.

Binary Maker 2.0 was designed to effectively accelerate this process by providing the user with a minimum of input parameters, offering various input modes to take advantage of what is already known about the binary, and allowing rapid changes of parameters with accompanying graphics which immediately indicate how those changes affected the preliminary light curve. The ability to see the light curve versus the data immediately greatly enhances the speed at which a preliminary solution is realized. This is especially true for spot modeling. ***Binary Maker 2.0*** originally was written from a need to draw spot sizes and positions on binaries (Bradstreet 1990), but has now evolved into a full-fledged light and radial velocity curve modeling program (Bradstreet 1991). A brief description of the program and several examples of its use follow.

Description of the Program

Binary Maker 2.0 is written in C and can be run on almost any machine which supports MSDOS 3.3 or higher, preferably 5.0 because of its improved memory management. Because of the nature of light curve programs, the computer needs as much speed as possible, preferably a 386- or 486-based device running at 33 MHz or higher with a math coprocessor. With enough speed, the component stars give the appearance of revolving in their orbits while the light and radial velocity curves are drawn. The program has been designed to be user-friendly, so that most functions are accessed through a pull-down menuing system (Adams 1990) and parameters are prompted for by the use of *input dialog* boxes, similar to those used on Macintosh machines. The menuing system is illustrated in Figure 1 where the **Run Menu** has been pulled down. Since these particular menu choices represent the major program options, they will each be briefly discussed below.

1. **Light curves** - This menu option initiates the main operation of the program, allowing

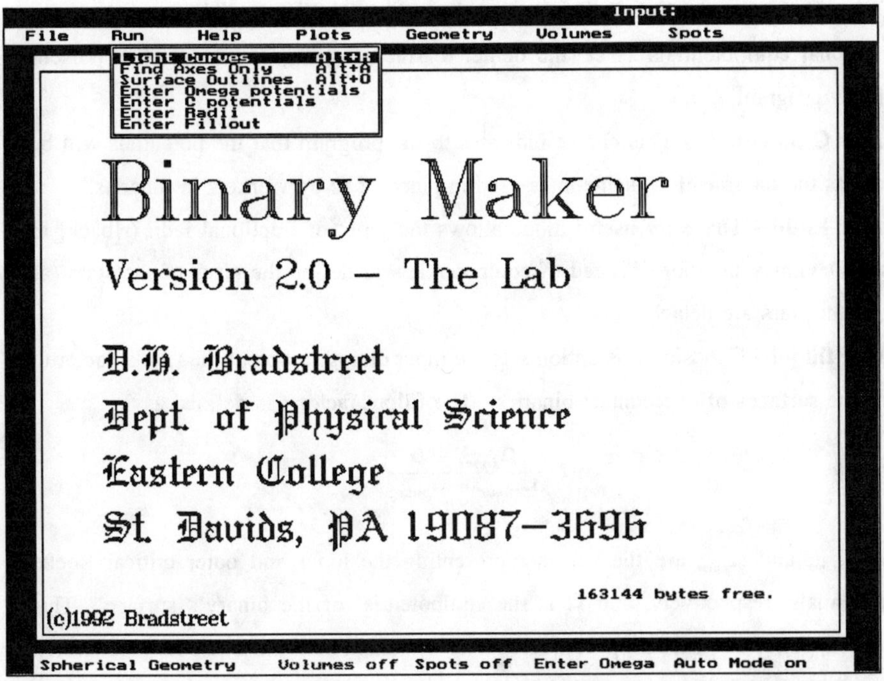

Fig. 1 - Title screen of Binary Maker 2.0 showing the menuing system

the user to input the necessary parameters to produce the three dimensional model of the binary as well as the light and radial velocity curves. Nearly any circular orbit close binary can be modeled, including detached, semi-detached, contact, and overcontact systems. Results can be saved as data files, screen image files, and PostScript files for output to a suitably equipped laser printer.

2. **Find axes only** - This quick calculation mode allows the user to immediately determine radii, equipotentials, fillout factors, etc., by simply entering the mass ratio and either equipotentials, fractional radii, or fillout factor for overcontact systems.

3. **Surface outline mode** - This feature constructs cross-sectional outlines (as seen from within the orbital plane) of the binaries and their inner and outer critical Roche equipotentials.

4. **Enter Omega potentials** - This menu choice sets the input dialog to accept the gravitational equipotentials as Ω (the default of the program) as used in the Wilson-Devinney program.

5. **Enter C potentials** - This choice indicates to the program that the potentials will be entered as the parameter **C**, sometimes used in place of Ω by workers in binaries.

6. **Enter Radii** - This very useful mode allows the input of fractional radii (r[back] in Wilson-Devinney notation) instead of equipotentials to define the sizes of the stars (as long as the stars are detached).

7. **Enter fillout** - Choosing this option sets the input dialog to accept the fillout factor to define the surfaces of overcontact binaries. The fillout factor f is defined as

$$f=\frac{\Omega_{inner}-\Omega}{\Omega_{inner}-\Omega_{outer}}$$

where Ω_{inner} and Ω_{outer} are the values representing the inner and outer critical Roche equipotentials, respectively, and Ω is the equipotential of the binary's surface. This feature expedites the modeling of overcontact systems since the equipotentials which produce a desired fillout are calculated and given as output.

The other menus are essentially ancillary to these functions. The **File Menu** allows the reading in of several different types of data files, namely, parameter files which include all necessary input to produce a given binary, files which contain up to 200 data points representing the observed light curve, and files which contain observed radial velocity measures. The data points which are read in will be displayed on the monitor as will be seen in the examples. The **Help Menu** provides on-line information concerning the use the program. The **Plots Menu** gives options to see previously made screens, output PostScript graphs, and also initiates the **Line profiles** mode which will be discussed later. The **Geometry Menu** selects between spherical (the program default) and cylindrical coordinates for constructing the surfaces of the stars. Some workers prefer cylindrical coordinates for overcontact binaries since the neck region is more accurately modeled by

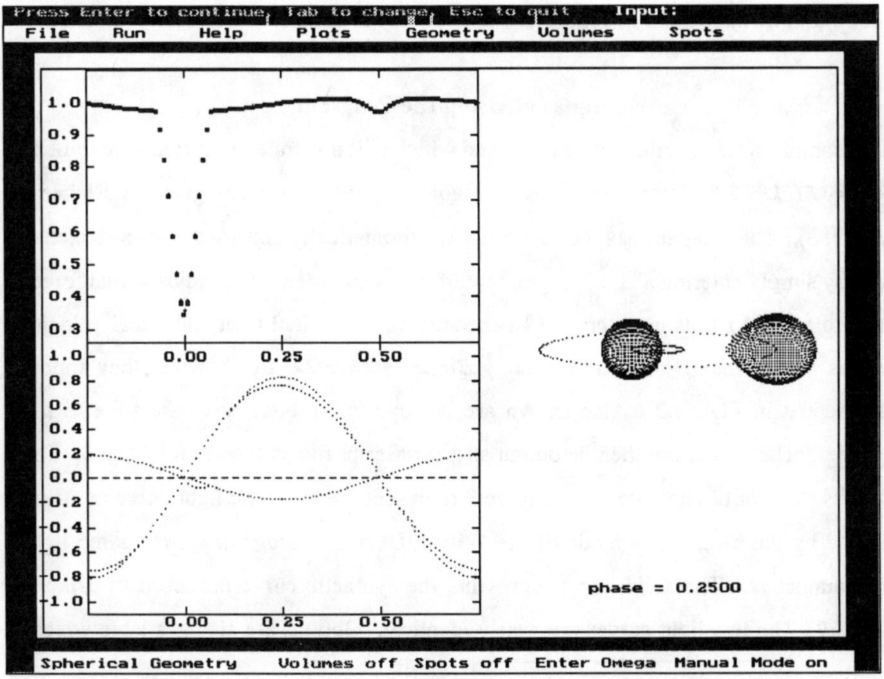

Fig. 2 - Appearance of the screen for the completed light and radial velocity curves of Algol.

that geometry. However, since so little radiation (at least at visible wavelengths) emanates from this region, the spherical coordinate representation of overcontact systems gives nearly identical results as using cylindrical. The **Volumes Menu** toggles the program into a mode by which volumes are calculated for each star by numerical integration and mean densities are calculated if the systems are contact or overcontact. Mean densities are important in stellar evolution studies. Finally, the **Spots Menu** prompts the user for all the appropriate spot parameters if spots are to be modeled. Wilson-Devinney spot notation (Yang and Wilson 1989) is used for spherical coordinates, Binnendijk notation (Binnendijk 1984) for cylindrical. The **Options Information Bar** at the bottom of the screen in Figure 1 indicates which options are currently invoked.

Examples of Using the Program

Several sample systems will now be discussed which will illustrate the various capabilities of ***Binary Maker 2.0***. First, let us look at Algol using the parameters given by Richards *et al.* (1988). The program has the capability to automatically construct a semi-detached system by simply entering a -1 when prompted for an equipotential or radius which forces the star to exactly fill its inner critical Roche surface. The final light and radial velocity curves as well as the 3D models created by ***Binary Maker 2.0*** are shown as they appear on the screen in Figure 2. Also shown are the optional orbital curves relative to the barycenter. These data can then be output as a PostScript file as shown in Figure 3. The plus signs ("+") in the light curve of Figure 3 represent the synthetic light curve of Algol generated by the black body mode of the Wilson-Devinney program for the same light curve parameters. The solid curve represents the synthetic curve produced by ***Binary Maker 2.0***. The two light curves are nearly identical. Options are also available which

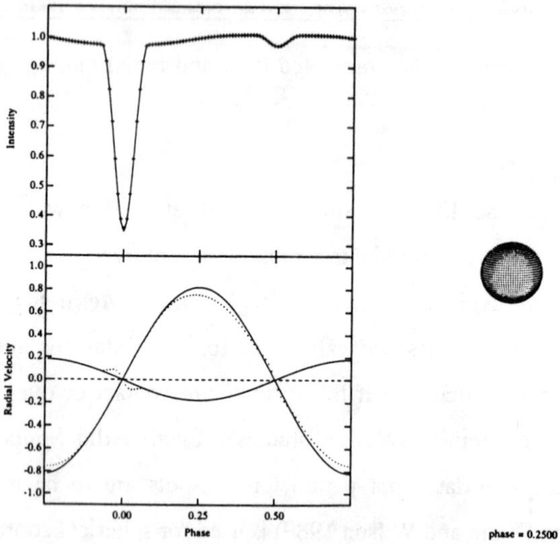

Fig. 3 - PostScript output of the completed Algol V light and radial velocity curve.

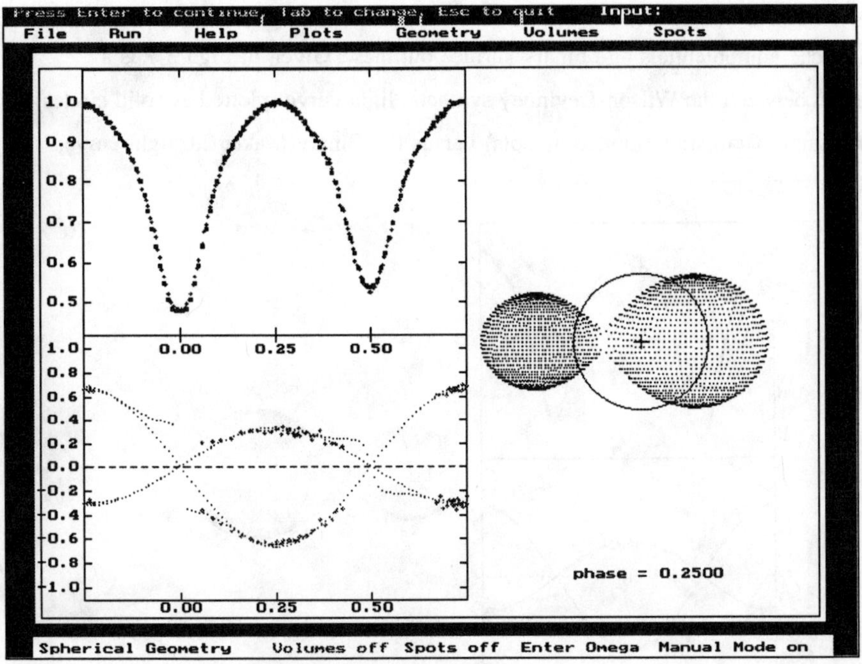

Fig. 4 - Completed screen display of AB And. The large circle represents the circumference of the Sun.

create separate full-screen versions of the light curve, radial velocity curve, or the three dimensional model of the binary. It is also possible to output these enlarged views as PostScript files.

The second example is AB And, a W UMa-type overcontact system. The final screen image of the system is shown in Figure 4, where the plotted *V* light curve normals were calculated from data given by Rigterink (1973) and the radial velocity data and light curve parameters were taken from Hrivnak (1988). The large circle plotted over the binary represents the circumference of the Sun relative to the component stars, another option of the program. The PostScript version of this screen is given in Figure 5. Figure

6 shows the results of using the **Surface Outline** option to produce the inner and outer critical Roche equipotentials and binary surface outlines. Given in Figure 7 is a comparison between the Wilson-Devinney synthetic light curve (plotted as solid circles) using the same parameters (plotted as dots) versus the Binary Maker 2.0 light curve,

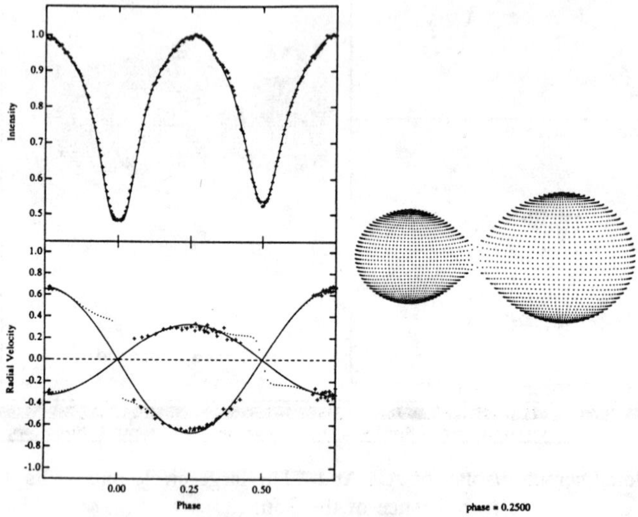

Fig. 5 - PostScript version of the screen display of AB And.

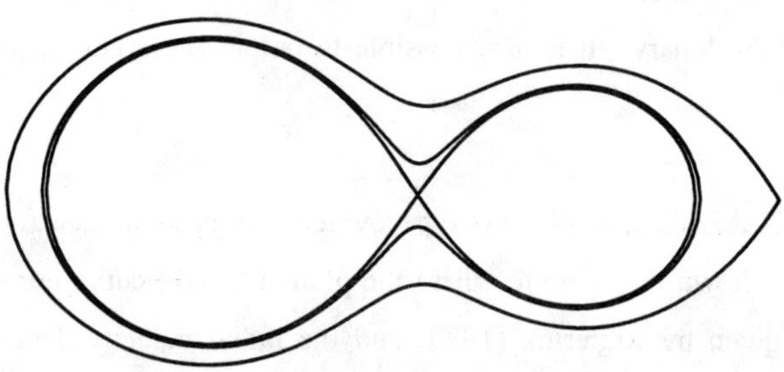

Fig. 6 - PostScript plot of the cross sectional outlines of the binaries and the inner and outer critical Roche surfaces for AB And.

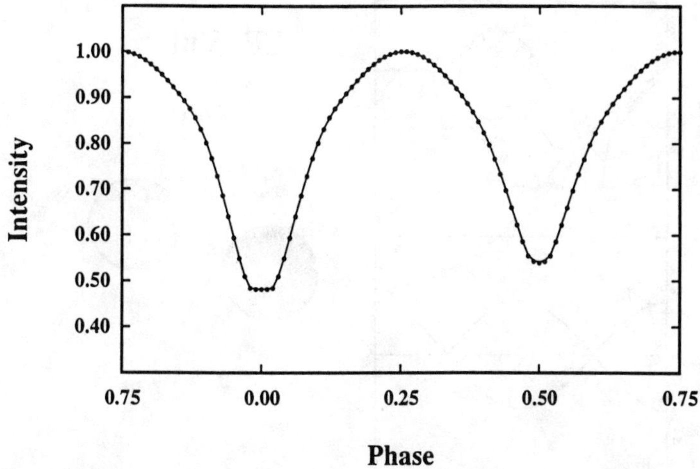

Fig. 7 - A comparison of the synthetic light curves of AB And as calculated by the Wilson-Devinney program (solid circles) and Binary Maker 2.0 (solid curve).

plotted as a smooth curve. Again the two synthetic curves are nearly identical, differing by only a few thousandths in intensity in the eclipses.

The third example is the highly active detached system ER Vul which has recently been studied by Guinan *et al.* (1991). Extensive multiwavelength (*IUE*) coverage of this system included several well covered *UBVRI* light curves. Figure 8 shows the PostScript output version containing the *V* light curve obtained by İbanoğlu *et al.* at Ege Observatory in Turkey in September 1991, the preliminary model fit (solid curve) by Binary Maker 2.0 (iterated by trial and error), and the radial velocity data of Hill *et al.* (1990). The fitting of all five light curves was made *much* easier by the ability to immediately see the

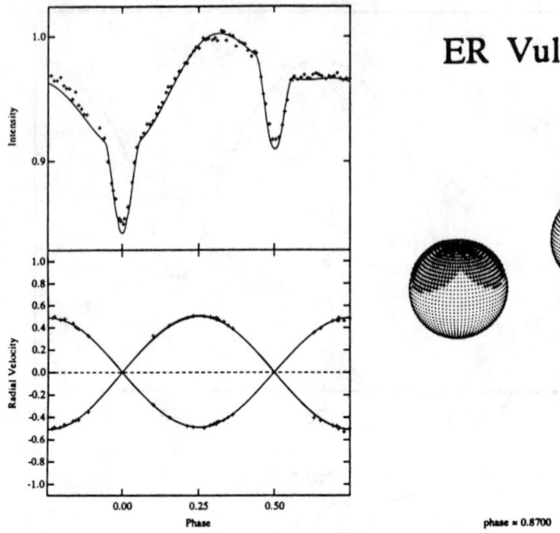

Fig 8. - PostScript output containing the September 1991 ER Vul *V* data of İbanoğlu *et al.*, the spot fit, and the radial velocity data from Hill *et al.* (1990).

effects of changing spot parameters with respect to the observed data. The goodness of fit for all five wavelengths (*UBVRI*) is essentially the same and was given by Guinan *et al.* (1991). The starspot parameters are given in Table 1.

The 1991 campaign to simultaneously observe ER Vul on the ground and with *IUE* resulted in numerous UV spectra around its 0.698 day orbit. One of the high dispersion long wavelength spectra (LWP) is shown below in Figure 9 centered near the λ2800 Å Mg II feature. The h and k lines of Mg II from each component are clearly seen demonstrating that both stars are sources of the emission. ***Binary Maker 2.0*** can construct synthetic line profiles by binning the intensities of the individual surface area elements according to their radial velocities (Binnendijk 1977; Rucinski 1985). The profiles show what would be expected *if* the source of absorption (or emission) is proportional to the intensity of the individual surface area elements binned in radial

ER Vul starspot parameters	spot #1	spot #2
colatitude	40°	40°
longitude	75°	190°
spot radius	35°	40°
temperature factor	0.93	0.90
star number	2	2

Table 1 - Starspot parameters for ER Vul for September 1991

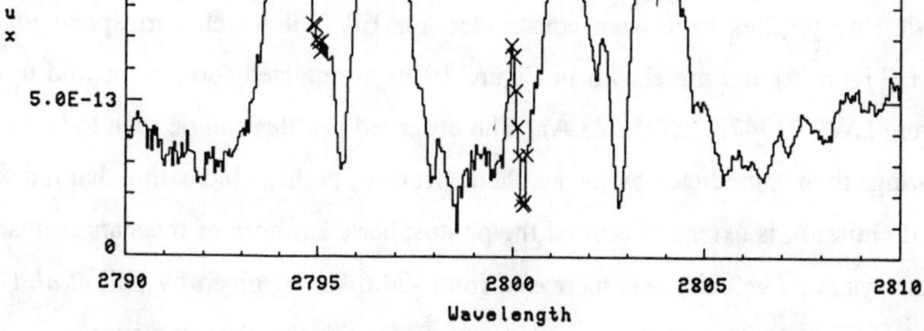

Fig. 9 - High dispersion IUE LWP spectrum of ER Vul obtained near 0.75P in September 1991. The Mg II h + k emission lines from each star are well resolved.

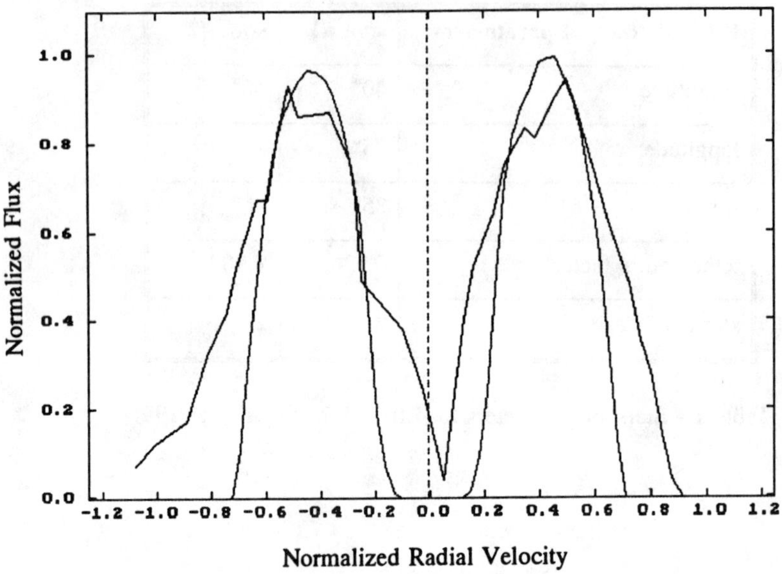

Fig. 10 - Synthetic line profile (narrow smooth curves) for ER Vul compared to the Mg II k line of LWP 21347H (broader jagged curves).

velocity space. The program then integrates the synthetic profile over a given phase interval to match the phase interval over which the exposure of the spectra was made. Synthetic line profiles have been constructed for ER Vul which correspond to LWP 21347H (Figure 9) and are shown in Figure 10 in normalized form compared to the k lines from LWP 21347H (2795.523 Å). The observed profiles can be seen to be broader in the wings than is predicted by the synthetic profiles, perhaps indicating that the source of Mg II emission is extended beyond the photospheric surfaces of the stars themselves. Also, the systemic velocity was increased from -24.6 km s^{-1} given by Hill *et al.* (1990) to -3 km s^{-1} in order to center the synthetic profiles with the observed ones.

The last example will again illustrate the **Line profile** option of the program using the W UMa binary VW Cep. This star has been studied intensively by Bradstreet and Guinan (1990) in both the visual and the ultraviolet using the *IUE*. The light curve of VW Cep varies fairly rapidly and the asymmetries have been modeled using large cool starspots on the larger cooler component of the system. Figure 11 shows the Strömgren y-band light curve (plus signs) obtained at Villanova University in September 1991. The smooth curve in the top of the figure represents the synthetic light curve which best fit the

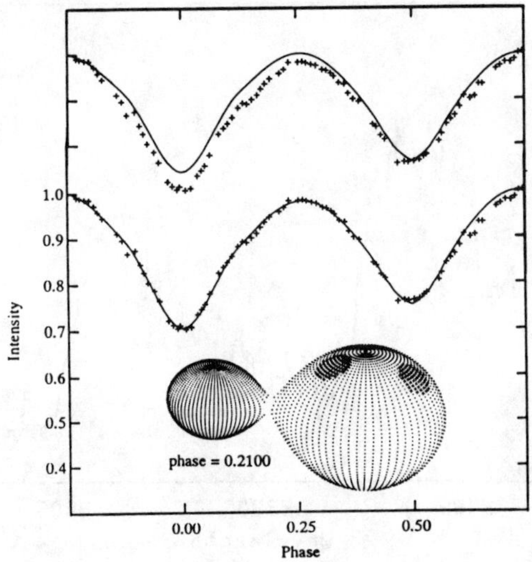

Fig. 11 - September 1991 visual light curve of VW Cep obtained at Villanova University and the unspotted (top curve) and spotted (bottom curve) synthetic light curves from ***Binary Maker 2.0***. The "unspotted" light curve fit the November 1990 data, also obtained at Villanova. The model of VW Cep with the two cool starspot regions is given in the bottom of the figure.

previous year's data obtained at Villanova in November 1990 when the system was apparently nearly "unspotted," or perhaps the spots were simply more uniformly distributed (Baluta *et al.* 1990). The bottom smooth curve is the ***Binary Maker 2.0*** preliminary fit with spots as shown in the three dimensional representation beneath it again compared to the September 1991 light curve. VW Cep was observed in September 1991 with the *IUE* nearly simultaneously with the ground based light curves. Among the many spectra obtained in this program were several high dispersion long wavelength (LWP) spectra. One of these spectra, (LWP 21306H) centered near the Mg II λ2800

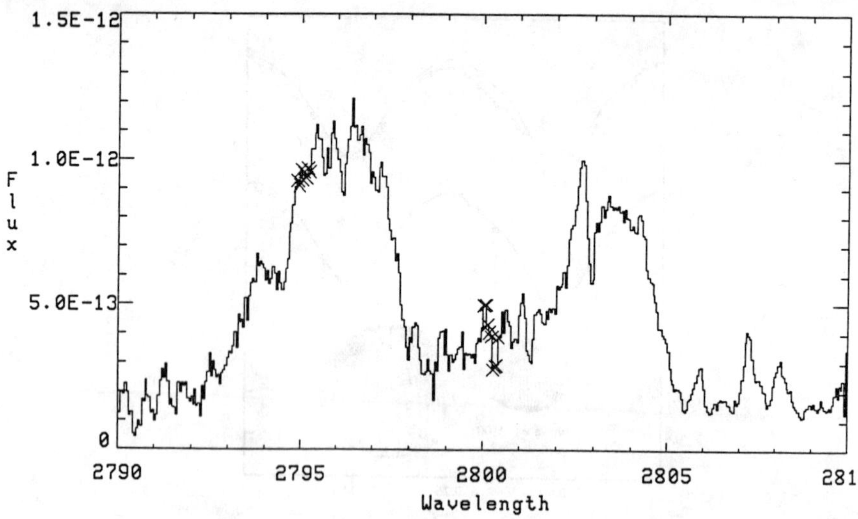

Fig. 12 - High dispersion LWP spectra of VW Cep centered near the Mg II λ2800 feature taken with the *IUE* in September 1991. This 90 minute exposure spanned phases 0.14P - 0.36P, and the emission profiles from both stars can be seen, especially in the k line near λ2795 Å.

feature, is shown in Figure 12. This spectra was exposed for 90 minutes covering phases 0.14P-0.36P. In Figure 12 the emission line profiles of *both* stars can be seen, the smaller profile of the less massive component discernible toward the shorter wavelength side of

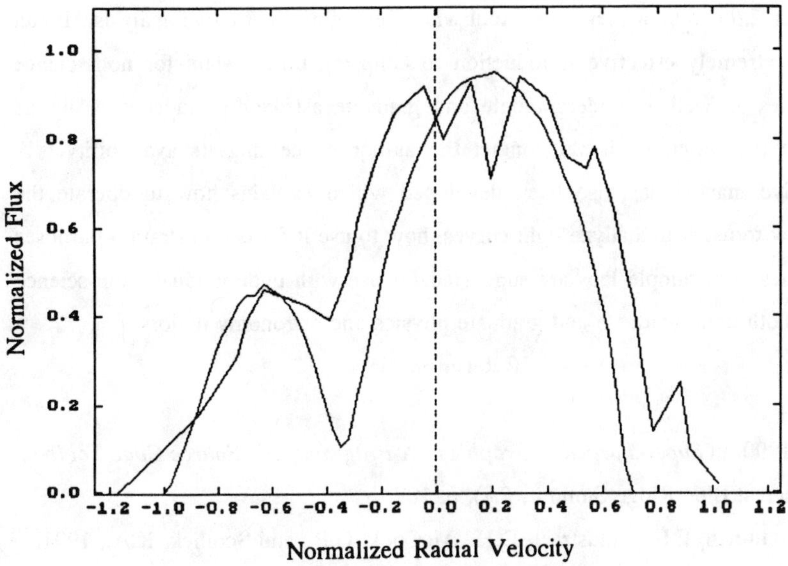

Fig. 13 - Synthetic line profile (smooth curve) for VW Cep compared to the Mg II k line of LWP 21306H (jagged curve).

the emission feature. Figure 13 shows a comparison between the theoretical line profile (smaller smooth curve) and the Mg II k line (larger "jagged" curve) for VW Cep corresponding to the same phase interval of LWP 21306H in Figure 12. As seen for ER Vul, the observed line profiles are broader than the theoretical profiles. It was also necessary to decrease the luminosity ratio L_2/L_1 from that found from the visual light curve (L_2/L_1 = 0.37) to match that of the observed profiles ($L_2/L_1 \approx 0.3$). These discrepancies may indicate an excess of UV flux emanating both from the connecting neck region and/or the more massive component as discussed by Bradstreet and Guinan (1990). Further analysis of all the spectra will hopefully lead to an explanation for the source of this apparent UV excess.

Binary Maker 2.0 can be a very useful tool when beginning light curve analysis. It can also be an extremely effective introduction to eclipsing binary stars for non-science undergraduates as well as undergraduate and graduate astronomy students. Anyone interested in the program should contact the author concerning its availability. A comprehensive manual has also been developed which explains how to operate the program, how to use it to analyze light curves, how to use it for demonstration purposes during lectures, and sample labs are suggested for use with undergraduate non-science majors and both undergraduate and graduate physics and astronomy majors.

References

Adams, L., 1990. in *Supercharged C Graphics - A Programmer's Source Code Toolbox*, (Windcrest:Blue Ridge Summit, PA), p. 107.

Baluta, C.J., Guinan, E.F., Bradstreet, D.H., McCook, G.P., and Scollick, K.A., 1991. *Bull. A.A.S.*, **23**, 835.

Binnendijk, L., 1977. *Vistas in Astronomy*, **21**, 359.

Binnendijk, L., 1984. *Publ. Astr. Soc. Pac.*, **96**, 646.

Bradstreet, D.H., 1990. *Bull. A.A.S.,* **22**, 1293.

Bradstreet, D.H., 1991. *Bull. A.A.S.,* **23**, 1414.

Bradstreet, D.H., and Guinan, E.F. 1990, in *Active Close Binaries*, NATO Advanced Studies Institute, ed. C. İbanoğlu. (Dordrecht: Kluwer), p. 467.

Guinan, E.F., Bradstreet, D.H., Etzel, P.B., İbanoğlu, C., Ready, C.J., Steelman, D.P., and Thrash, T.A., 1991, *Bull. A.A.S.*, **23**, 1412.

Hill, G., Fisher, W.A., and Holmgren, D., 1990. *Astron. Astrophys.*, **238**, 145.

Hrivnak, B.J., 1988. *Ap. J.* **335**, 319.

Kang, Y.W., and Wilson, R.E., 1989. *A. J.*, **97**, 848.

Richards, M.T., Mochnacki, S.W., and Bolton, C.T., 1988. *A. J.*, **96**, 326.

Rucinski, S.M., 1985. "Contact Binaries: Observational Properties", in *Interacting Binary Stars*, eds. J.E. Pringle and R.A. Wade. (Cambridge), p 85.

Wilson, R.E., 1979. *Ap. J.*, 234, 1054.

Wilson, R.E., and Devinney, E.J., 1971. *Ap. J.*, **166**, 605.

Problems Arising in the Modeling of the W Ser-Type Binary RX Cas*

Il-Seong Nha
Yonsei University Observatory, Seoul 120-749, Korea
Ho-Il Kim
Sobaeksan Astronomical Observatory, KAO, Daejeon 305-348, Korea

ABSTRACT. *UBV* observations of RX Cas were made at Yonsei University Observatory in one observing season from Sept. 1988 to Feb. 1989 for 66 nights, which cover six consecutive orbital cycles. In this observing season the RX Cas system has undergone light changes from the *QUIET* state (cycle Nos. 1 and 2) to the *VIOLENT* state (cycle Nos. 3 to 6). Significant dips are clearly shown on light curves of both sides of the primary eclipse at its *QUIET* state which may be interpreted as a strong indication of the presence of an active, optically thick disk around the mass gaining hot secondary component of this system.

SHORT BACKGROUND STORY

With a large number of observations of RX Cas (V_{max} = 8.6v, Sp = A + K1 III, P = 32 days) in ten years, Kalv(1979) deduced a 516 day period of long term light variations in three different stages, so-called MIN, MED, and MAX, according to the light levels. Kriz *et al.*(1980) and Fernandes(1985) independently confirmed these light variations, but they are contradictory to each other for the periodicity of long term variations by Kalv. There have been a number of light curve solutions reported recently(Kalv 1979, Strupat 1987, Andersen *et al.* 1989).

Despite all these efforts, there are still difficulties in modeling light curves which combine observations from many cycles. This is because details of the peculiar cycle-to-cycle variations over RX Cas' long orbital period merely get buried in the light curve scatter.

UBV observations of RX Cas were made in 1988-89 season as a part of the *Ten-year Observing Program for Long Period Eclipsing Binary Stars* (1982 - 1992) at Yonsei

---

* Yonsei University Observatory Contribution No. 96.

University Observatory(YUO). A total of 412 observations (V = 158, B = 158, U = 96) covered six consecutive orbital cycles in about 200 days suffice to make light curves homogeneously covered in phase without much of its complications. The standardized indivisual observations are tabulated in Table 1.

In this paper we present light curves of RX Cas, discuss the nature of photometric behaviors, and raise the problems involved in the modeling of RX Cas.

QUITE AND VIOLENT NATURES

UBV light curves according to the Julian dates are presented in Figure 1. We notice two

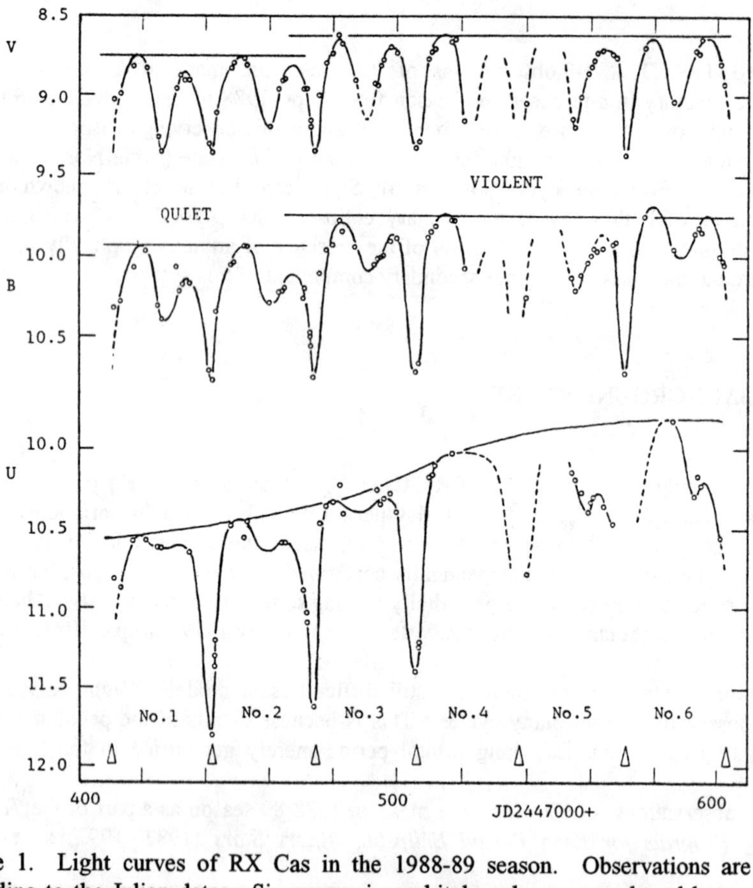

Figure 1. Light curves of RX Cas in the 1988-89 season. Observations are arranged according to the Julian dates. Six successive orbital cycles are numbered between arrows which indicate times of minimum. Free-hand curves are shown for each color.

Table 1. Standardized observations of RX Cas

HJD	V	B-V	U-B	HJD	V	B-V	U-B
47411.1938	9.073	1.315	0.447	47443.1088	9.128	1.210	1.032
47411.2521	9.024	1.296	0.501	47443.1189	9.143	1.212	
47411.3237	9.031	1.265	0.520	47443.1202	9.142	1.211	
47413.2471	8.999	1.313	0.604	47443.1222	9.145	1.218	
47413.2582	8.987	1.290	0.596	47443.1662	9.093	1.246	
47413.2689	8.991	1.286	0.560	47443.1678	9.099	1.245	
47417.1576	8.826	1.252	0.479	47443.1760	9.114	1.237	
47417.2327	8.838	1.249	0.525	47443.1775	9.116	1.225	
47421.1276	8.806	1.155	0.678	47443.2037	9.114	1.225	
47421.1387	8.825	1.144	0.631	47443.2110	9.100	1.287	
47421.1795	8.846	1.149	0.511	47443.2579	9.111	1.288	0.906
47421.2373	8.868	1.118	0.573	47443.2684	9.120	1.268	0.818
47425.3357	9.269	1.052	0.307	47443.2777	9.114	1.286	
47426.1154	9.347	1.048	0.190	47443.2796	9.116	1.292	
47426.1253	9.337	1.052	0.224	47443.3150	9.102	1.224	
47426.1632	9.387	1.021	0.229	47443.3164	9.105	1.231	
47426.1745	9.363	1.029	0.253	47443.3231	9.099	1.241	
47426.2330	9.382	1.045		47443.3246	9.095	1.248	
47426.2408	9.370	1.047		47443.3315	9.100	1.250	
47431.1654	8.959	1.117		47443.3329	9.098	1.260	
47432.1881	8.871	1.342		47448.1597	8.833	1.236	0.412
47432.1932	8.888	1.354		47448.1699	8.829	1.231	0.421
47433.1416	8.963	1.292	0.471	47449.2935	8.799	1.246	
47433.1515	8.848	1.272		47452.1354	8.778	1.167	0.629
47435.1266	8.908	1.269	0.471	47452.1459	8.770	1.171	0.602
47435.1390	8.912	1.266	0.465	47453.2208	8.819	1.126	0.518
47441.2562	9.308	1.415		47460.1621	9.197	1.105	
47441.2648	9.309	1.403		47463.2414	8.966	1.285	
47441.2661	9.310	1.383		47463.2487	8.965	1.270	
47441.2677	9.312	1.361		47464.2432	8.925	1.318	0.316
47441.2746	9.313	1.354		47464.2550	8.889	1.320	0.419
47441.2759	9.310	1.364		47465.2361	8.886	1.300	0.408
47441.2822	9.328	1.387		47465.2453	8.906	1.307	
47441.3104	9.289	1.460		47471.1056	8.953	1.320	0.613
47441.3120	9.294	1.458		47471.1167	8.965	1.319	0.631
47441.3201	9.306	1.484		47471.1290	8.964	1.320	0.615
47441.3219	9.306	1.492		47471.1998	8.948	1.314	0.586
47442.0410	9.364	1.423		47472.1929	8.973	1.242	0.591
47442.1168	9.367	1.411		47472.2067	8.942	1.256	0.628
47442.1589	9.363	1.413	1.019	47473.0909	9.157	1.326	0.548

Table 1. *Continued.*

HJD	V	B−V	U−B	HJD	V	B−V	U−B
47473.1036	9.171	1.342	0.573	47518.1152	8.655	1.112	
47473.1579	9.200	1.367		47521.0647	9.159	0.938	
47474.1554	9.345	1.410	0.865	47521.0733	9.161	0.908	
47474.2346	9.355	1.405	0.856	47540.0751	9.050	1.213	0.520
47476.1717	8.903	1.212	0.355	47540.0857	9.044	1.225	0.524
47478.1850	8.803	1.165	0.395	47554.0520	9.155	1.003	0.020
47478.1958	8.792	1.176	0.368	47554.0619	9.152	0.967	0.007
47480.1862	8.741	1.197	0.420	47555.0201	9.199	1.017	-0.036
47480.1978	8.727	1.200	0.408	47555.0281	9.198	1.007	0.010
47482.0890	8.622	1.160	0.459	47557.0542	9.007	1.108	0.170
47482.0998	8.630	1.158	0.436	47557.0639	9.008	1.118	0.149
47483.1499	8.671	1.155	0.561	47559.0239	8.855	1.180	0.337
47483.1617	8.693	1.174	0.573	47559.0343	8.872	1.173	0.354
47487.1366	8.911	1.134		47559.9947	8.810	1.169	0.446
47493.1724	8.948	1.120		47560.0063	8.831	1.203	0.308
47493.1795	8.918	1.133		47561.0495	8.778	1.178	0.365
47494.0788	8.835	1.178	0.283	47561.9942	8.732	1.240	0.323
47494.1491	8.836	1.191	0.198	47564.0000	8.715	1.237	0.421
47495.1261	8.820	1.193	0.335	47564.0092	8.713	1.245	0.400
47496.1060	8.777	1.222	0.324	47567.0815	8.736	1.189	0.545
47498.1236	8.697	1.208	0.391	47568.0091	8.766	1.158	
47498.1334	8.707	1.213	0.351	47568.0163	8.757	1.153	
47500.1226	8.737	1.260	0.417	47570.9950	9.381	1.342	
47500.1317	8.745	1.237	0.408	47571.0029	9.353	1.400	
47506.1428	9.340	1.409	0.616	47576.9774	8.710	1.090	
47506.1519	9.331	1.390	0.715	47576.9831	8.676	1.029	
47506.1891	9.322	1.392		47585.9740	9.031	0.934	-0.176
47506.1959	9.347	1.409		47585.9844	9.013	0.927	-0.083
47506.2025	9.333	1.370		47585.9950	9.059	0.901	-0.124
47507.1458	9.289	1.368	0.595	47592.9920	8.712	1.145	0.439
47507.1558	9.340	1.384	0.493	47593.9882	8.811	1.035	0.327
47510.1346	8.777	1.106	0.299	47594.0006	8.666	1.132	0.370
47510.1442	8.760	1.131	0.284	47595.9933	8.644	1.203	0.387
47511.1406	8.730	1.119	0.320	47600.9711	8.863	1.194	0.334
47512.1181	8.707	1.114	0.320	47600.9841	8.741	1.190	0.451
47512.1269	8.713	1.103	0.270	47601.9875	8.892	1.139	0.663
47517.2100	8.670	1.130	0.237	47601.9976	8.957	1.082	
47517.2179	8.657	1.124		47602.0061	8.986	1.153	
47518.1088	8.676	1.133		47602.0138	8.904	1.149	

significant different states of light levels, *QUIET* and *VIOLENT*, in *V* and *B* which are indicated with straight lines for the outside eclipse phases as shown in this figure. The differences between the states are $0.^m13$ and $0.^m18$ for *V* and *B*, respectively. There seems no trace of continuous development between the states, which suggests that the transition time would be too short, at least for *V* and *B* passbands, to be noticed between cycles Nos. 2 and 3 during our monitoring. The light levels of the primary minimum have remained constant in *V* and *B* at 9.37 and 10.76, respectively, while the secondary minimum light tends to change every cycle in both colors.

However, the *U* curve has, on the other hand, behaved completely different way from the *V* and *B*. The *VIOLENT* state has been gradually developed from the *QUIET* state for both outside maxima and the primary minima, and the secondary minima are extremely complicated changes in depth, shape, and even position from cycle to cycle. *U* light curves of the first two cycles, Nos. 1 and 2, are in agreement marginally, but those of cycles Nos. 3 and 4 are in the transition with rapid light increase between the first and the last cycles. The magnitude difference outside eclipse between the two extremes is about 0.72.

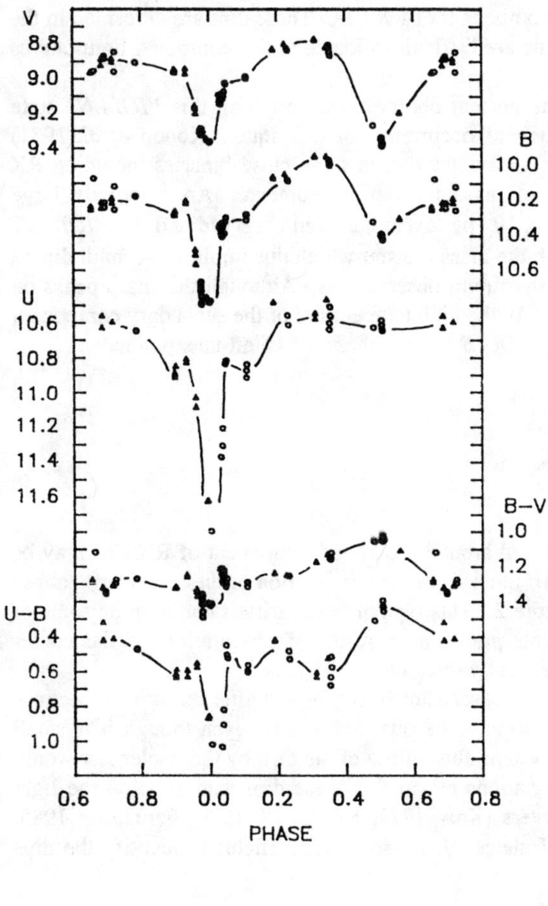

Figure 2. Light and color curves of RX Cas. Symbols used are open circles for the cycle No. 1 (JD2447411 - 2447443) and filled triangles for cycle No. 2 (JD2447448 - 2447474). Solid curves are free-hand curves.

Orbital period of RX Cas has continued lengthening even after the new period of 32.33010 days by Andersen *at al.*(1989), and thus for short-term phase computation within the range of JD2445178 and hopefully for near future the following linear ephemeris is made;

$$Min.\ I = HJD2445598.911 + 32.^{d}336959E.$$

Light curves were made using this ephemeris for the phase computations for each cycle individually and then compared them to see if there were curves match one another. But, as was clear in the Figure 1, there found no such curves except those of *QUIET* state, cycle Nos. 1 and 2, which are shown in Figure 2. Free-hand curves connecting data points show the shapes similar to those previously known, in general, but with fine structure. Significant dips appear on both sides of the primary eclipse at about $+0.^{p}11$ from the mid-eclipse, and another one at phase $0.^{p}24$. All of these are color dependent, deeper with shorter wavelength. In spite of the disappearance of the latter dip in *B-V* and *U-B* color curves, the former two dips at $+0.^{p}11$ are there clearly, indicating strong evidence of an accretion disk formed around the hot component of RX Cas. These dips are deformed in the light curves of *VIOLENT* state and thus are difficult to locate in the combined light curves for cycles, Nos. 3-6.

However, it is not clear from the present observations how long this *VIOLENT* state lasts and what would be the frequency of occurrence of this state. Kondo *et al.*(1981) discovered a substantial excess of far-ultraviolet flux in three close binaries including RX Cas and interpreted it as the result of mass transfer phenomena. An extremely large increase of orbital period, $\Delta P/P = 6.3 \times 10^{-7}$ per cycle has been observed and the *VIOLENT* state of RX Cas may be correlated with the mass transfer which this implies. A small dip on phase 0.24 needs its confirmation by the future observations. Although the sharp peaks on phase 0.5 in *B-V* and *U-B* support strongly the high temperature of the secondary component (Plavec and Koch 1978), a great deficit of *U-B* around phase 0.3 is left unexplained.

EVIDENCE OF ACCRETION DISK

The existence of an accretion disk formed around the hotter component of RX Cas may be supported by several evidences. One of them is the longer duration of the secondary eclipse than that of the primary as seen in Figure 2. This type of peculiarity is common on some of W Ser-type binaries and has been interpreted as a result of obscuration of the cooler component by the optically and geometrically thick disk.

The appearance of dips at the light curves shortly before and after the primary eclipse is clear in Figure 2. If the disk were to have its own luminosity, even though it is small fraction to the total luminosity of the system, the eclipse of the disk by the cooler star would affect the light levels. Careful examinations reveal that these dips also exist on the light curves of RX Cas by previous observers (Kalv 1979, Kriz *et al.* 1980, Fernandes 1985, Andersen *et al.* 1989) at their *QUIET* states. It is, however, difficult to identify the dips during the *VIOLENT* states.

Collisions between the mass ejected from the cooler star and the disk would produce bright region at a particular phase, which makes the light curves outside eclipse asymmetry. Our light curves, in fact, have more light at $0.^{p}25$ than at $0.^{p}75$ by $0.^{m}12$ and $0.^{m}23$ for *V* and *B*, respectively. This indicates that the collisions occur on the hemisphere which is shown around $0.^{p}25$.

According to the above evidences of the disk of RX Cas, we presume that the depressed regions at phases 0.90 (-36°) and 0.10 (+36°) would represent the disk size and radius. The phase angle(θ) of the semi-duration of the disk eclipse have a relation with the radius of the cooler component(r_c) and orbital inclination(i) as

$$sin\ \theta = (r_d + r_c)\ sin\ i,$$

where r_d represents the radius of disk in the unit of separation between the two stars. We adopt the values of Andersen *et al.*(1989) 0.28 and 80° for r_c and i, respectively. Since is around 36° as shown in Figure 2, r_d becomes about 0.3. This value, however, is almost double the value determined by Andersen *et al.*(1989) but small enough to be inside of Roche critical lobe calculated by mass ratio of Andersen *et al.* for RX Cas.

ACKNOWLEDGEMENT

We are indebted to many student observers of Yonsei University who took part in the observations. One of us (ISN) is grateful to Dr. F. B. Wood who allowed his Card Catalog to be used freely along with generous hospitality during a visit to University of Florida, and to Dr. K.-C. Leung for his warm hospitality during ISN's visit to University of Nebraska, where the final version of this work was made. This work was supported financially in part by a Yonsei University Faculty Research Grant in 1990, for which ISN expresses his gratitude.

REFERENCES

Andersen, J., Pavlovsky, K., and Piirola, V. 1989, A&A, 215, 272.
Fernandes, M. 1985, BAV Rundbrief, 34(2), 49.
Kalv, P. 1979, Tartu Astr. Obs., Teated No. 58, 1.
Kondo, Y., McCluskey, G. R., and Wu, C. C. 1981, ApJS, 47, 333.
Kriz, S., Arsenijevic, J., Grygar, J., Harmanec, P., Horn, J., Koubsky, P., Pavlovsky, K., Zverko, J., and Zdarsky, F. 1980, Bull. Astr. Inst. Czech., 31, 284.
Plavec, M. and Koch, R. H. 1978, IBVS, 1482.
Strupat, W. 1987, A&A, 185, 150.

Spot Solutions for Two RS CVn Systems: ER Vul and V711 Tau

D.S. Zhai, R.X. Zhang, and X.B. Zhang
Beijing Observatory, Chinese Academy of Sciences, Beijing 100080, China

Abstract

ABSTRACT: Based on spot model and Roche approximation for binary stars, the photometric analysis was carried out for the BV light curves of both the eclipsing system ER Vul and the non-eclipsing V711 Tau. The preferable spot parameters were obtained by using differential correction adjustment and least squares technique.

1 Introduction

For modelling the distorted light curves of RS CVn systems, the spot assumption has been widely accepted by the most authors while there is a considerable range in the sophistication of model. The spot effect was also taken into consideration by the well-known Wilson-Devinney synthetic light curve program in its edition of 1978. Recently, Zhai et al.(1988) and Kang and Wilson (1989) extended the function of W-D program to include differential correction of spot parameters. Compared with others, their method has the following physical characteristics:

1. It is based on Roche model for binary system and has taken into account the photometric proximity effects due to tides and irradiation between two components;
2. Spot parameters can be adjusted by the differential correction with numerical derivatives. The results are of impersonal values with probable errors;
3. Multi-color light curves can be solved simultaneously for spot parameters.

In this paper the general procedures introduced by Zhai et al. (1988) was followed in dealing with the light curves of ER Vul and V711 Tau. First, by inspecting the behaviour of the O-C curves of a temporarily average solution without spots, it was found that two spots are needed for each system. The preliminary longitude of each spot can be estimated directly from those O-C curves, too. The probable latitude of spots was determined by a grid method introduced by Zhai et al.(1988).At each assumed latitude the parameters of Roche and spot model were alternatively adjusted with the differential corrections for the best fitting of observations. As concerns the non-eclipsing system V711 Tau, strong physical constraints have been added to the Roche model based on spectroscopic data.

2 The Results

1. ER Vul:

ER Vul is a short period (0.6981 days) RS CVn system and also a double-lined spectroscopic binary with two G-type components G0V + G5V. Its light curves are so contaminated by complex spot effects that the secondary eclipse has greater depth than the primary at times (e.g., see Northcott and Bakos, 1967 or Evren, 1988). Budding and Zeilik (1987) published a solution for the 1978 light curves with a model of a set of completely dark spots at 45° latitude.

Here, the 1957 BV light curves of Northcott and Bakos were employed to perform our new photometric analysis with a two-spot model. The original mass ratio q=m_2/m_1=0.93 was taken from the radial velocity solution (c.f. Batten, 1989). The final convergent solution parameters are given for the Roche model in table 1 and the spot parameters in Table 2. In figure 1, the theoretical light curves are plotted together with observations. It is shown by our results that two spots, one on the primary component and the other on the secondary, can reconstruct the main appearance of the observed light curve but the spot on the secondary may have a temperature a little higher than that of photosphere.

2. V711 Tau (=HR1099):

V711 Tau is a very noteworthy object (P=2.8377 days, Sp.:G5V-IV + K1I V) for its strong chromospheric activities. In a recent paper Kang and Wilson (1989) made a unified analysis of a total of nine light curves published from 1963 to 1981. The migration period of the spot on V711 Tau was found from fairly good determination of spot longitudes. In this paper we present the new photometric analysis of the recent observations obtained just after a great optical flare in 1989 (Zhang and Zhai et al., 1990).

For the non-eclipsing detached system V711 Tau, the additional constraints on the Roche model have to be added for reliable spot

Table 1. Solutions of Roche model parameters for ER Vul and V711 Tau

Star	ER Vul (1957)		V711 Tau (1989)	
	5500 Å	4500 Å	5500 Å	4500 Å
$L_1/(L_1+L_2)$	0.544± 0.015	0.548± 0.015	0.226±0.027	0.252±0.028
$L_3/(L_1+L_2+L_3)$	---	---	0.0639	0.0582
x_1	0.64	0.77	0.65	0.74
x_2	0.66	0.81	0.83	0.91
i	66.60± 0.20		33.00	
$q=m_2/m_1$	0.9337		1.25	
$g_1=g_2$	0.32		0.32	
$A_1=A_2$	0.50		0.50	
	Comp.1	Comp.2	comp.1	Comp.2
T	6000 K	5904± 38 K	5370 K	4623
Ω	4.570 ± 0.083	4.602 ± 0.091	9.04 ±0.72	4.73 ± 0.12
r_{pole}	0.2726± 0.0059	0.2593± 0.0058	0.128±0.012	0.320±0.010
r_{point}	0.2904± 0.0075	0.2753± 0.0073	0.129±0.013	0.355±0.014
r_{side}	0.2781± 0.0064	0.2641± 0.0062	0.129±0.012	0.330±0.011
r_{back}	0.2859± 0.0071	0.2714± 0.0069	0.129±0.012	0.344±0.013
Ω_{inn}	3.6414		4.1462	

Table 2. Spot parameters for ER Vul and V711 Tau

Satr	ER Vul		V711 Tau	
	Spot 1 (Comp.1)	Spot 2 (Comp.2)	Spot 1 (Comp.2)	Spot 2 (Comp.2)
Lat.	55.0± 1.4	63.5± 3.9	60.68± 0.31	75.5± 4.9
Long.	214.94± 0.50	224.2± 1.3	261.53± 0.75	61.9± 3.1
R_s	13.70± 0.35	22.3± 1.5	20.02± 0.65	9.48± 0.70
T_s/T_{photo}	0.892± 0.004	1.026± 0.002	0.837± 0.006	0.936± 0.022
$S_s/S_{hem.}$	2.85%	7.46%	6.04%	1.37%

analysis. We set $q=m_2/m_1=1.25$ and $i=33^o$ in accord with Fekel (1983). The temperature T_1= 5370 K and T_2=4623 K are adopted from the calibration (Jager and Nieuwenhuijzen, 1987) with the spectral type of each component. As concerns the surface potentials Ω_1 and Ω_2, there are two choices: one is to let them freely adjust. It turned out that $r_1=0.251$ and r_2/r_1= 1.37 are close to those obtained by Kang and Wilson (1989), but quite different from the estimation by Fekel who gave a ratio r_2/r_1 = 3.0 from his rotation measures. The other way is to put additional constraint on Ω_1 and Ω_2, referring to Fekel's data, too.

It was found that the Ω_2 is not highly correlated with Ω_1 within a large rang of given Ω_1 values, a comprehensive fact that the sub-

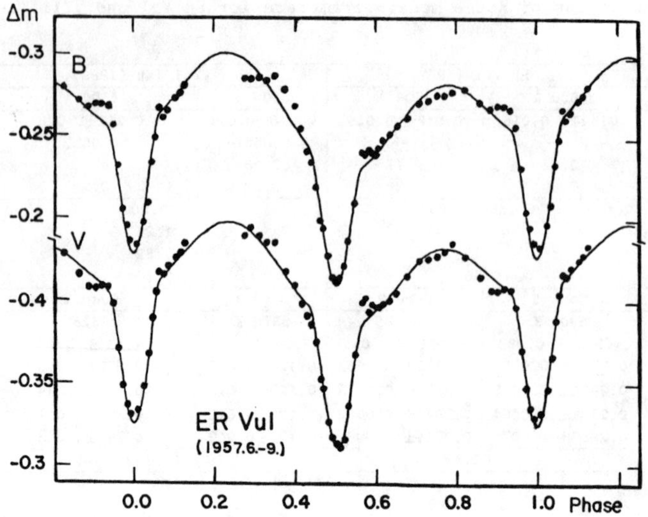

Figure 1. Light curve of ER Vul in 1957.

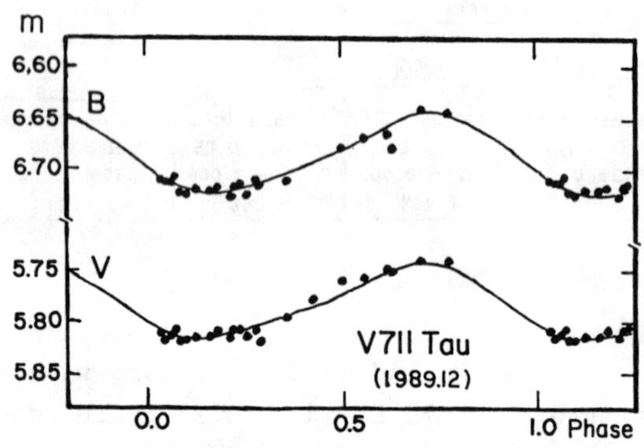

Figure 2. Light curve of V711 Tau in 1989.

giant component 2 is the spotted star and also the main source of photometric effects. Starting from a larger Ω_1, a second convergent solution was obtained at Ω_1=9.037 and Ω_2=4.735. The side radii of the Roche lobes turned out to be r_1=0.129 and r_2=0.330 with a ratio r_2/r_1=2.558. If the magnitude differences between two components in the UBV system as estimated by Fekel (c.f. Henry and Hall, 1991) are taken into consideration, the ratio of radii of two assumed black bodies will be about 2.0. So our solution 2 would be preferable except for the loss of some fitting accuracy in light curves. However, the spot parameters both in solution 1 and 2 are fairly close to each other. Table 2 lists solution 2 as our final result and Figure 2 shows the theoretical light curves as well as the observations.

3 Conclusions and Discussions

1. The spot assumption as per our approach has simulated well the distorted light curves for both ER Vul and V711 Tau even though each of them has specially photometric complexity. However, the solution for each system is but one of the most probable solutions resulting from our differential corrections adjustments; that is to say, it is a stable instead of an unique solution. They are limited by the observational errors as well as the model itself. For their small orbital inclination such like that of the non-eclipsing binary system V711 Tau, the solution could not get rid of the pole spots within the latitude less than $90° - i$ from the pole. These spots were kept in sight of observers at any orbital phase and gave influence mainly on the systematic brightness rather than on the shape of light curves.

2. The photometric proximity due to the "reflection" and ellipticity effects was found to be about 0.03 mag, nearly one third of the amplitude of light curves (see Figure 3). This is a non-negligible complexity for one to derive a reliable spot solution. It is believed that the spherical-star model will not be a good chioce for the RS CVn system such like V711 Tau.

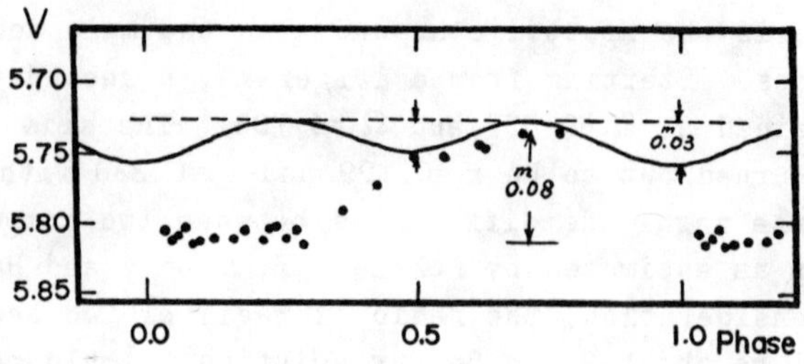

Figure 3. Photometric proximity effects for V711 Tau.

3. It was found by W-C program that the "phase shift" of the light curves of V711 Tau was quite different in different wave band. The light minimum of the system was about 0.07 and 0.10 phase later in U than in B and V band, respectively. This seems unlikely to be caused by the accidental errors. The further study, together with R and I observations is completely needed.

ACKNOWLEDGEMENTS:This research was supported by the Natural Science Foundation of China. The authors would like also to express their thanks to Beijing Observatory for providing the necessary computer time.

Reference:

Batten,A.H.,Fletcher,J.M.,and MacCarthy,D.G.:1989, Eighth Cataloque of the Orbital Elements of Spectroscopic Binary Systems,, P.112.

Budding,E.,and Zeilik,M.:1987, Astrophys.J.,**319**,827.

Evren,S.:1988, Astrophys.Space Sci.,**140**,125

Fekel,F.C.:1983, Astrophys.J.,**268**,274.

Henry,G.W.,and Hall,D.S.:1991, Astrophys.J.,**373**,L9-L12.

De Jager,C.,and Nieuwenhuijzen,H.:1987, Astron.Astrophys.,**177**,217.

Kang,Y.W.,and Wilson,R.E.:1989, Astron.J.,**97**,848.

Northcott,R.J.,and Bakos,G.A.:1967, Astron.J.,**72**,89.

Zhai,D.S.,and Chen,H.L.:1988, Acta Astrophys.Sin.,8,265. (see:1989,Chin.Astron.Astrophys.,**13**,9.)

Zhang,R.X.,Zhai,D.S.,Zhang,X.B.,Zhang,J.T.,and Li,Q.S.:1990, Inf.Bull.Var.Stars, No.3456.

The Information Limit Optimisation Technique

Timothy Banks
Physics Department, Victoria University of Wellington,
P.O. Box 600, Wellington, New Zealand

ABSTRACT:

It is important not to over, or under, parameterise a data set during analysis, as otherwise physically unreliable values will result. This problem is of considerable concern in the analysis of eclipsing binary light curves, where it is always tempting to include the effects of more "accurate" formulations, or of additional, physical processes. The Information Limit Optimisation Technique (ILOT) seeks to avoid this by examining the χ^2 parameter space about the adopted solution, to ensure that the information content of the data is matched by that of the model being applied to it. New developments in the ILOT technique, which is currently enjoying a growth in usage, are also outlined. Solutions derived by the ILOT approach have recently been shown to be comparable to those found by the other main computer methods of eclipsing binary light curve analysis.

The basic problem of modelling any data with theory lies in constraining the applied models, while only introducing a minimum of assumptions. It is important not to over-parameterise the model for a given data set, which will undoubtedly result in a closer, but rather imprudent, fit to the data. For instance, a parent sinusoidal distribution can provide information on three meaningful parameters only - wavelength, amplitude, and phase offset. The application to a real data set of a model fitting a larger number of parameters than these could certainly appear to fit noisy data better, but the physical interpretation of the resulting parameters is then somewhat inappropriate.

The Information Limit Optimisation Technique (ILOT) attempts to ascertain the "information limit" of a data set by checking a solution's determinacy using the Hessian matrix (second order derivatives of χ^2 with respect to the various parameters being optimised; see Bevington, 1969, for details) in order to ensure that too many parameters are not regarded as determinable in the analysis. The method examines the χ^2 parameter space about the adopted solution, which should be a strict minimum, with closed contours of χ^2 about it (see Figure 1). Banks and Budding (1990a) demonstrated how the accuracy of individual parameters are resolved at deteriorates as the set of sought parameters is increased in number, until one or more of the parameters are free to vary over some range between their physical limits. Such a breakdown in solution determinancy is indicated in the inverted Hessian ("error") matrix as an effectively zero gradient (along some range of a parameter axis defined by the final parameter "step sizes" used to estimate the gradients numerically). The ILOT technique regards indeterminacy as providing a clear limit on the number of parameters that can be derived by the fitting model for a given data set. Naturally, the number of derivable parameters is dependent on the quality of the data being analysed. In addition, inspection of the Hessian matrix, and in particular its eigenvalues and eigenvectors, permits a useful insight into parameter interdependence, along with an estimate of the solution's formal accuracy in a given parameter.

The ILOT method's fitting function for eclipsing binaries is of a relatively simple and compact algebraic form, allowing the modelling programs to be run on even personal computers with an acceptable run time. Kopal's (1959) original alpha integrals are constructed using the recursion relations given by Budding (1973). The components' original surface fluxes, as approximated from model atmosphere data, as well as the radiative ("reflection") and gravitational ("ellipticity" and "gravity brightening") interactions are all accounted for. Three standard optimisation techniques are employed in an iterative approach by the program - parabolic interpolation in a "step by step" (single parameter) and conjugate directions ("Powell") modes, and a multi-parameter "vector" mode - to search for improvements in the χ^2 goodness of fit of the model light curve to the data as selected parameters are judiciously varied.

The program switches automatically between these methods depending on the convergence rate and certain user-specified parameters (for details see Banks, 1989; and Adby and Dempster, 1974, for a more general background on optimisation methods). The ILOT method has been found (see e.g. Banks *et al.*, 1990; Banks and Budding, 1989, 1991a; Banks, 1991a) to produce effectively the same geometric parameters as the Wilson-Devinney (1971), WINK (Wood, 1971), and Light (Hill, 1979) programs, over a range of separations, including contact binaries with the Wilson-Devinney and Light approaches (WINK being only suitable for well detached systems), and light curves exhibiting distortions. There does appear to be some systematic difference between the fractional luminosities derived by the ILOT and Wilson-Devinney methods, which relates to whether the effective temperature which generates a given relative luminosity can be regarded as a more basic parameter than an isolated empirical relative luminosity value itself, given that blocking factors can still intervene between the assigned temperature and a derived luminosity. If a detailed model atmosphere is to be included, then the choice of effective temperature as the principal parameter would be appropriate. Hence, the ILOT codes can estimate at least as good parameter sets as the other main computer model methods for eclipsing binaries.

This raises the point that since the same physical processes are being studied, various analysis techniques should converge towards the same 'picture' despite their procedural differences. Formulations of the problems differ, and so it is not surprising that the nature of the parameters used to describe the physics do as well. A case in point is how stellar radii are treated in different models - with the point and side radii of the Wilson-Devinney (1971) model compared to, say, the ILOT spherical radius, which is that the star would possess if it was removed from the binary into isolation. A meaningful comparison between models therefore requires an understanding on what their parameters physically represent. Such comparisons are desirable, and indeed part of scientific method, allowing independent confirmations of a suggested scenario for a system, particularly given the increasing complexity of the analysis programs themselves, with the corresponding problems in testing the actual code itself.

Light curves exhibiting other distortions have been modelled by the ILOT technique in a three stage approach. Firstly the light curve is processed for standard eclipsing binary effects (e.g. Nield, 1987; Zeilik *et al*, 1988; Zeilik *et al*, 1990). The residuals from this "fit" are then interpreted in terms of circular dark spots (see e.g. Budding and Zeilik, 1987) associated with stellar activity, or a "hot spot" created by a mass transfer stream's impact (see e.g. Edalati *et al.*, 1991; Banks *et al.* , 1990, Banks and Sullivan, 1990; Forbes, 1990). The original data is then "corrected" for the best spot model's effect, and re-analysed, taking into account only the expected binary system effects (see Figure 2). It is assumed that the two light curves (binary and spot) are independent, and so superposable. The underlying ILOT philosophy is applied to the spot analysis (whose algebraic fitting function is taken from Budding, 1977a, and may include spot bands or penumbrae).

The ILOT technique has been applied, with different fitting functions, to the luminosity distribution of globular clusters, binary colour curves (Walkington, 1988), radial velocity curves (see Banks *et al.* , 1990), light curves of pulsating variables (Budding, 1992, 1977b), and the H_β index (Edalati *et al.* , 1991).

The ILOT approach is currently being applied to, and developed for, a variety of situations, and provides a reasonable package of programs that can be used in the analysis of eclipsing binary light curves. It is a robust technique which allows the information content of the data to be matched, while allowing the data to be neither under or over - parameterized. Such a feature has been rather neglected in light curve analysis, for example a 22 free parameter model has been applied by another group to a photometric data set that the ILOT approach could only confidently extract five parameters from (Banks and Budding, 1990b). Such an "overly" accurate model unfortunately does nothing but lull the investigator into a false sense of security and confidence.

References:

Adby, D.R., and Dempster, M.A.H., 1974, *Introduction to Optimisation Techniques*, Chapman and Hall.

Banks, T., 1989, *Unpublished M.Sc Thesis* , Victoria University of Wellington.

Banks, T., 1991a, in *Proceedings of the Third New Zealand Photoelectric Photometry Conference* , 129, Eds: J. Richard and E. Budding, Fairborn Press.

Banks, T., 1991b, in *Proceedings of the Astronomical Society of Australia,* **9(1)**, 148.

Banks, T., and Budding, E., 1990a, *Astrophys. Spa. Sci.* , **167**, 221.

Banks, T., and Budding, E., 1990b, *Earth, Moon, and Planets* , **49**, 15.

Banks, T., and Budding, E., 1991, *The Observatory,* **111**, 38.

Banks, T., and Sullivan, D.J., 1990, *IBVS # 3415* .

Banks, T., Sullivan, D.J., and Budding, E., 1990, *Astrophys. Spa. Sci.,* **173**, 77.

Bevington, P.R., 1969, *Data Reduction & Error Analysis for the Physical Sciences*, McGraw-Hill.

Budding, E., 1973, *Astrophys. Spa. Sci.* , **22**, 87.

Budding, E., 1977a, *Astrophys. Spa. Sci.* , **46**, 407.

Budding, E., 1977b, *Astrophys. Spa. Sci.* , **48**, 249.

Budding, E., 1992, *Basic Astronomical Photometry* , submitted, Cambridge University Press.

Budding, E., and Zeilik, M., 1987, *Astrophys. J.* , **319**, 827.

Edalati, M.T., Banks, T., and Budding, E., 1992, in *IAU Symposium 151*, in press.

Forbes, M.C., 1990, *Unpublished M.Sc Thesis*, Victoria University of Wellington.

Hill, G., 1979, *Publ. Dominion Astrophys. Obs.* , **15**, 297.

Nield, K.M., 1987, *Unpublished M.Sc Thesis*, Waikato University, New Zealand.

Walkington, M., 1988, *Personal Communication.*

Wilson, R.E., and Devinney, E.J., 1971, *Astrophys. J.* , **166**, 605.

Wood, D.B., 1971, *Astron J.* , **76**, 701.

Zeilik, M., De Blasi, C., Rhodes, M., and Budding, E., 1988, *Astrophys. J.*, **332**, 293.

Zeilik, M., Ledlow, M., Rhodes, M., and Budding, E., 1990, in *Astron. Soc. Pacific Conf. Series,* **9**, 252.

Figure (1) :

χ^2 Contours against spot latitudes for a two spot solution for the single chromospherically active star AB Dor assuming an inclination of 70 degrees. The minimum indicates that the first spot prefers latitude ≈ 0.35 radians, while the second spot prefers below 0.65 radians. Only one hemisphere is plotted here. The opposite hemisphere is somewhat similar, with the minimum being surrounded by higher values of χ^2 . This diagram shows a cross section through the χ^2 hyper- surface, and demonstrates how a determinate solution is a minimum in this surface.

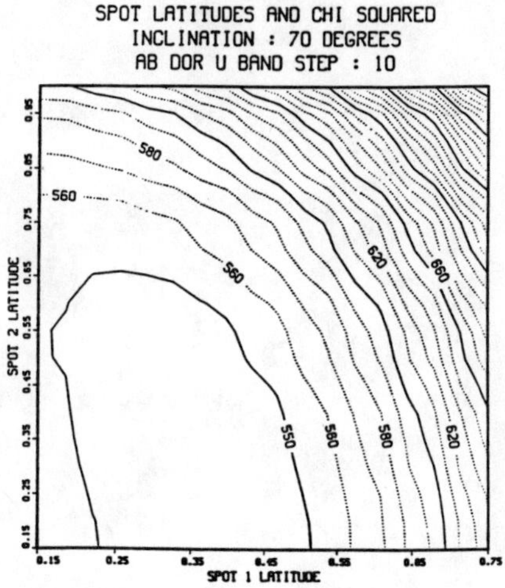

Figure (2) :

A B band light curve for RT Andromedae after the effect of a 'star spot' has been accounted for. Details of this ILOT analysis may be found in Banks (1991a).

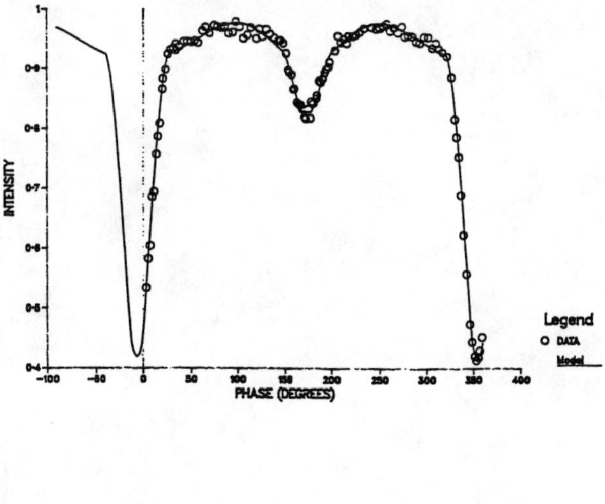

Comments on Light Curve Modelling Techniques and a Review of a Poster Paper

E. Budding
Carter Observatory, P.O. Box 2909, Wellington, New Zealand

Abstract

A review of a poster paper not published elsewhere in this volume, orally reviewed along with several others for the light curve modelling meetings held at the XXIst IAU General Assembly in Buenos Aires and in Cordoba, is presented, and is preceded by remarks made at the time of the review.

Introduction - The general curve fitting problem

In a general way we can regard the curve-fitting problem as how to make an appropriate transformation from information presented in a *data space* to a suitable representation in *parameter space*. The parameters appear in a *fitting function*, whose main variables are those of the data space.

In a typical practical situation the data appear as a set of discrete points, having a certain non-uniform, probabalistic distribution. Different sets of parameters map different versions of the fitting function, overlapping different values in data space, so that the net probability of such versions will vary in response to the chosen parameter set. We can consider an *optimal* solution to the curve-fitting problem to be provided by a parameter set which maximises this probability. It can be shown that if data ordinates are normally distributed about their mean at each abscissal value, for example, that this maximisation is equivalent to the minimisation of X^2, formed by the sum of the differences between observed and calculated ordinates divided by the variances of the normal distributions at such abscissae.

The parameters of the fitting function for 'light curves' are generally not free to take any real value. The data is also finite in extent, implying a corresponding limitation on their information content, or, equivalently, the number of independent parameter values, *m* say, in the set $\mathbf{a}_j$ to be specified by the curve-fitting solution. We thus face a constrained minimisation problem, with a built-in determinacy issue. It can be shown that, as we increase *m*, then uncertainty factors $\Delta \mathbf{a}_j$ (simply related to the expected errors of the specifications of the $\mathbf{a}_j$) increase progressively, until, at some point, (the 'information limit') such factors, for one or more $\mathbf{a}_j$, formally $\rightarrow \infty$; the solution becomes indeterminate, or non-unique. Put simply, a finite data sample can only yield a finite amount of information. These issues figure in five poster papers which were orally reviewed in Cordoba, four of which are published in this volume. The remaining poster paper, presented at the IAU but not submitted for inclusion here, is reviewed.

'BX Peg --- Simultaneous Photometric and Spectroscopic Solution' by R. G. Samec and D. P. Hube

The authors use the 1983 version of the Wilson Devinney (WD) code which has the option for simultaneous light and radial velocity curve fittings. The advantages of this approach can be simply seen in information terms as: more information in $\rightarrow$ more information out. Since we know that some parameters must be the same for both fitting functions, *e.g.*, the mass ratio, the *combined* parameter set specification is fewer than for two *separate* fittings, so that the uncertainty factors are correspondingly reduced.

The point can be made that contact binary light curves can be expected to be not as intrinsically informative as more detached binaries. This can be visualized in Fourier analysis terms, if we can argue that the light curves are of quasi-sinusoidal, smoothly varying shape. The number of Fourier coefficients required to adequately represent such a shape is perhaps not too many. Nevertheless, the contact constraint implies that the mass ratio q, filling factor f and inclination i are the only major physical parameters to be specified, so that the data are likely to be quite adequate for this. The authors find that the combined spectroscopic and photometric solution gives effectively the same q as before, while i and f are both reduced --- but very slightly. Confidence intervals were not specified for the solutions, but my impression is that the difference between the new and old solutions is not statistically significant. A final interesting physical point is made in the paper that BX Peg appears to be steadily shortening its period, and thus perhaps progressing towards the 'coalescence' that certain evolution scenarios predict for these binaries.

Acknowledgements

Thanks go to Dr. E. Milone for arranging this interesting meeting, and inviting me to make the reviews.

Reference

Wilson, R. E., and Devinney, E. J., 1971, *Astrophys. J.*, **166**, 605.

Some Comments on Light Curve Solution of Eclipsing Binaries

A.M. Cherepashchuk
Moscow State University, Moscow, Russia

The problem of the light curve solution of eclipsing binary systems is a typical inverse problem that in the general case is ill-posed, *i.e.*, small errors lead to large deviations in the solution. A stable solution of the problem needs knowledge of so-called "a priori" physical information about the solution. The algorithms of the solving of ill-posed problems based on application of "a priori" information (so-called regularisation algorithms) have been developed by A.N. Tikhonov. These regularisation algorithms have been applied to light curve solutions for eclipsing binary systems with extended atmospheres (see, for example, Cherepashchuk 1975). The basic assumption here is the following: the functions of interest describing the distribution of the brightness and absorption properties over the disk of the star with the extended atmosphere are monotone and non-negative ones. Application of this method to the interpretation of multi-wavelength light curves of eclipsing Wolf-Rayet binary V 444 Cyg allows us to find the basic characteristics of Wolf-Rayet star and its extended atmosphere (see Cherepashchuk, Eaton, and Khaliullin, 1984).

In the most typical case when both components of the eclipsing binary are stars with thin atmospheres a great amount of "a priori" information about the properties of the stars permits the consideration of the inverse problem of the light curve solution as a parametric problem. If the number of the parameters of interest is limited, it is a well-posed problem that may be solved by well known light curve synthesis methods (see, for example Wilson and Devinney 1971 and later improvements). Parametric models for X-ray binary systems and SS 433 object have been developed using light curve synthesis methods by our group (see, for example, Antokhina and Cherepashchuk, 1987). The algorithms for the determination of the values of the parameters and their confidence intervals using statistical criteria (Fisher,hi**2) have been developed for the light curve solution of X-ray binaries and SS 433. A very important problem for light curve solution in the framework of parametric model is the comparison of the model and observations using statistical criteria. It should be stressed that the adopting or rejecting the model according to a statistical criterion does not prove the reality or nonacceptance of the model. Four possibilities must be considered here:

1. The model is real and is adopted according to the statistical criteria;
2. The model is not real but is adopted according to the criteria (so called error of the second kind);
3. The model is real but is rejected according to the criteria (the error of the first kind);
4. The model is not real and is rejected according to the criteria. Because it is impossible to prove that the model is real, there are non zero probabilities for all four cases. That is why we adopt the model according to the statistical criteria because we have small ground to reject it and we must keep in mind that there is some non-zero probability that this model is not real. That is why the problem of estimate of the confidence intervals (errors) for the parameters of the model is very related to the problem of statistical demonstration of the model.

All these problems are investigated in the following three books:
Goncharsky, Cherepashchuk, and Yagola 1978; 1985); and Goncharsky, Romanov, Cherepashchuk 1991)[1].

---

[1]Dr. Cherepashchuk is providing an English-language summary of some of the previously untranslated material discussed in these books for a compendium of light curve analysis methods by Kallrath and Milone. - Ed.

References

Antokhina, E. A., and Cherepashchuk, A. M. 1987. *Astron. Zh.* 64, 562.

Cherepashchuk, A. M. 1975. *Soviet Astron.* **19**, 47.

Cherepashchuk, A. M., Eaton, J. A., and Khaliullin, Kh. F. 1984. *Astrophys.J.* **281**, 774.

Goncharsky, A. V., Cherepashchuk, A. M., and Yagola, A. G. 1978. *Numerical Methods for the Solving of Inverse Problems of Astrophysics*, "Nauka", Moscow, (in Russian).

Goncharsky, A. V., Cherepashchuk, A. M., and Yagola, A. G. 1985. *Ill-Posed Problems of Astrophysics*, "Nauka", Moscow, (in Russian).

Goncharsky, A. V., Romanov, S. Yu., Cherepashchuk, A. M. 1991. *Limit-Number Parametric Inverse Problems of Astrophysics*, "Moscow University Press", Moscow, 1991 (in Russian).

Wilson, R. E., and Devinney, E. J. 1971. *Astrophys. J.* **166**, 605.

Discussion
Comments by D. Popper

1. As a followup on Ed Budding's statement that his student has compiled a list of light curve synthesis programs, I suggest that a useful outcome of this meeting might be a concerted effort, comparing results of applying those programs in active use to one or more suitable light curves. A difficult region in which considerable work is being done now is systems with unequal components having radiative envelopes with periods in the 0.4-0.7 days range (*e.g.* V1010 Oph, GR Tau. *etc.*). Contact systems of the W UMa type have their configurations too closely constrained for critical tests, it seems to me.

2. In my view, temperature is not an appropriate parameter in solutions of light curves. Surface flux ratios and light ratios in the wavelength of observation are the quantities that enter into the light variation directly. Information should be made available to the interested reader that would facilitate his own evaluation of absolute surface fluxes and/or effective temperatures, independent of the temperatures that may be assigned or drop out of the light curve solution. After a light curve solution has been completed by whatever means, with all the adopted, non-radiative parameters held fixed, the light curve in each wavelength band should be solved for the flux ratio and/or light ratio. Then, anyone interested in evaluating absolute fluxes, temperatures, *etc.*, will have the relevant information in hand. This procedure will be of optimum value only if the wavelength bands employed by the observer are tied to a standard system observationally or are clearly defined by narrow band filters.

Concluding Remarks

The papers in this volume provide a window on the status of light curve modeling by a vital segment of the eclipsing binary star modeling community.

On examining the list of contents, one might be attempted to conclude that the Wilson-Devinney model has been given excess attention. However, this is not the case. In the Commission 42 report for 1991, percentages of light curve solutions that were carried out with different techniques are tabulated for triennial reporting periods beginning with 1975-78. The results, summarized in Fig. 1, demonstrate that the Wilson-Devinney program is at present the synthesis and analysis code of choice for most of the community. It is not surprising, therefore, that half the papers in this volume involve the use of one version or another of the Wilson-Devinney program. On the other hand, Fig. 1 shows that the 6 light curve methods or codes cited do not constitute 100% of the total in any triennium. In the interest of simplicity, our graph's cutoff has been placed at ≥7%, so programs with smaller usages do not appear. Some of these programs, such as Hill's LIGHT2, or Kallrath's simplex code, both of which are described in the current volume, show increasing usage with time. Hill's code was used in 5-6% of published modeling papers in the 1984-7 and 1987-90 triennia, and it is likely to undergo further increase unless its features are absorbed by better known codes. Philosophical concerns about the nature of modeling radiative properties of the stars - especially when these are complicated by circumstellar material - will continue to give value to alternate methods and code for the foreseeable future. This point has been raised by Ed Budding, among others. In particular, highly accurate results are obtainable with the EBOP program discussed by Paul Etzel, if an eclipsing binary's geometric properties alone are desired. For undistorted systems, especially eccentric orbit systems, the results with this code are unexcelled, although as a

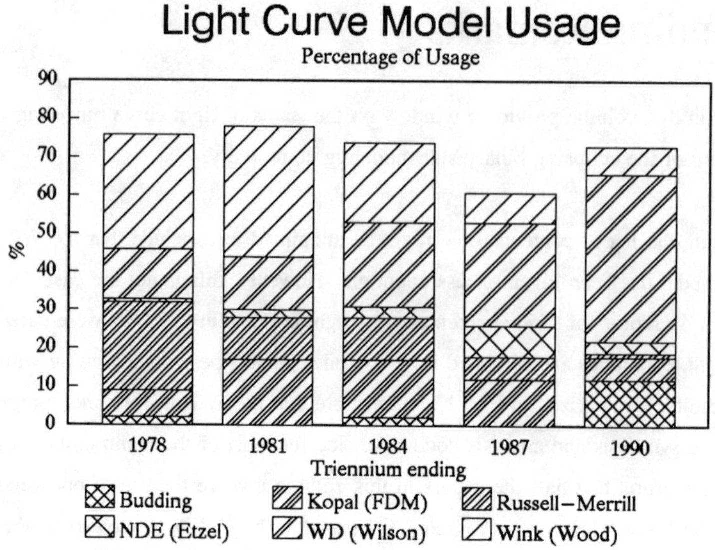

Figure 1. Summary of Triennial Usage Data (from Wilson 1990).

recent series of tests (Milone, Stagg, and Kurucz 1991, cited in the Stagg and Milone paper in this volume) shows, the final results, under identical assumptions, are compatible with those from their upgraded version of WD.

An area which merits more discussion than it has received thus far in these proceedings is that of comparative modeling with different codes, although occasional comparisons are mentioned in Commission 42 Program Notes and triennial reports. As Professor Popper has suggested, it would be valuable to compare the optimized modeling results from different programs and procedures. In this volume, Hill refers to such comparisons, but they are not reproduced or cited; Stagg and Milone refer to comparative work done on AI Phe, but the detailed discussion appears elsewhere (Milone *et al.* 1992). Accordingly, to provide at least one such set, the properties of the relatively shallow contact system TY Bootis is provided in Table 1. This test is not really definitive for several reasons, but it is interesting nonetheless, because the same data (that of Carr 1972) have been analyzed by three methods, and another data set (Samec's) by two. The *WD* results are from the work of Milone *et al.* (1991), those of *Light2* are from Rainger, Hilditch, and Bell (1990), while the frequency-domain (*FD*) and Russell-Merrill (*RM*) solutions are from Niarchos (1978) and Carr (1972), respectively. The Milone *et al. WD* results are based on radial velocity data, new, high precision photometric observations by D. H. Bradstreet, but are tested against observations by Samec and Bookmyer (1987), Carr, and RAO (University of Calgary data) observations; the results of those tests, which were followed to convergence in the same way as the adopted results involving the Bradstreet and RAO data, appear in the *second* row for each parameter). Differences among the derived elements for these three data sets are not significantly different (as determined by 3σ differences among formal errors) except for ι, ΔT, and L_1; for ι the data agree within 1%, for L_1, within 6%. The difference among the derived T_2 values are larger (15% from the Carr data solution) and may indicate temporal and localized variations, which, outside of eclipse, might well be modeled by spotted areas. In Table 1, we compare our adopted solution (using Bradstreet's data) to the Rainger *et al.*, Samec, Carr, and Niarchos solutions. The former two use Samec's data, the latter three, Carr's. The Bradstreet and RAO data required spots to fit the light curve adequately. These spots were optimized, one parameter at a time and checks were made of the results on the system and other component parameters as well as other spot values after each optimization. The procedure is similar to that carried out by Lepasset *et al.* (1992) involving the optimisation of the mass ratio q, where

Table 1. TY Bootis Modeling Results Compared

Parameter	WD[a]	WD[b]	Light2[c]	FD[d]	RM[e]
a ($R_\odot$)	2.32(1)	2.32	2.15(2)	...	...
		2.32(2)			2.32(2)
e[f]	0	0	0	...	0
i (deg.)	77.5(1)	77.1(1)	76.6(2)	70.8(7)	...
		77.0(1)			78.3(1)
q (m_2/m_1)	2.138(7)	2.24(2)	2.29(4)	(0.23)	...
		2.125(7)			2.153(10)
$\Omega_{1,2}$	5.370(8)	2.746(6)	...	...	...
		5.374(7)			5.374(16)
T_1^f (K)	5834(150)	5623	6180(200)	...	...
		5502			5524
ΔT (K)	365(5)	323(7)	385(12)	...	...
		332(6)			310(5)
$\langle R_1 \rangle$ ($R_\odot$)	0.71(1)	0.73	0.69(1)	...	...
$\langle R_2 \rangle$ ($R_\odot$)	1.02(1)	1.06	1.00(1)	...	
k (R_s/R_g)	0.70(2)	0.69	0.69(2)	0.48	0.94
$L_1/(L_1+L_2)_B$	0.430(1)	0.403(2)	0.433	0.202	...
		0.411(1)			
$L_1/(L_1+L_2)_V$	0.417(1)	0.388(1)	...	0.213	...
		0.409(1)			
f	0.12(3)	0.10	0.13(2)	...	...
V_γ (km/s)	-38.7(7)	...	-41.9(23)	...	...
spots	2	0	0	0	0

[a] Wilson-Devinney (as modified by EFM & CRS) applied by Milone *et al.* (1991);
[b] Wilson-Devinney applied by Samec *et al.* (1989). Modeling includes preliminary RV data of Milone *et al.* 1987. $\Omega_{1,2}$ value reflects the procedure of adding $0^{P}.5$ to all data, equivalent to inverting stars 1 and 2 so that $T_2 > T_1$. No spots were assumed. Except for Ω, the parameters have been converted to W-type notation; [c] *Light2* applied by Rainger *et al.* 1990, with RV data. T_2 (in *WD* notation) was fixed at 5800K and T_1 found to be 6180K. RVs were fit by sine curves and not modeled with the light curves. V_γ computed separately for stars 1 and 2. The mean is given; [d] Frequency Domain applied by Niarchos (1978) to Carr's data. Averages of the separate B and V light curve solutions are shown.
[e] Russell-Merrill method applied by Carr (1972) to his B and V data only; [f] fixed.

all values but q were kept constant and adjusting q from run to run until a minimum of the resulting parabolic distribution of SSRs could be found. Samec *et al.* (1989) also attempted a *WD* solution based on his own light curve and on only preliminary radial velocity data (Milone *et al.* 1987). The Rainger *et al.* radial velocity data are fewer but have somewhat less scatter than the radial velocity data of Milone *et al.*, so that their input should sharpen the determination of the mass ratio, when modeled along with the photometric data. Niarchos (1978) analyzed the UBV data of Carr (1972), but neither had access to radial velocity data, which may account for their clearly erroneous results. Modeling experiments performed without the use of RV data provide useful tests of uniqueness determination techniques (for example, ambiguity between A- and W-type solutions clearly have arisen in the absence of definitive information from radial velocity curves), but spectacular failures provide the object lesson that in the absence of such data, a thorough exploration of parameter space is essential. For the systems studies by Kallrath *et al.* (1992), the simplex method certainly helps in such an exploration - provided that parameters are not forced into their constraint limits too early in the iteration process. In any case, to conclude our discussion on TY Boo, infrared data obtained at the Mt. Lemmon IRTF and KPNO, additional optical data from Odell, and Rainger *et al.*'s radial velocity data will be analyzed to improve the determined quantities, and investigate the sources of light curve asymmetries.

We clearly have not exhausted the subject. In particular, the state of light curve modeling of exotic objects as dwarf novae, symbiotic systems, and polars is not well-represented here, although the important modeling elements of disks, streams, and winds, and for the first time, the polarization synthesis properties are to be incorporated into general light curve modeling code. The paper of Stagg and Milone cite some of work done in the more specialized areas. The increasing accessiblity of infrared, microwave, and even radio observations on the one hand, and the far UV and x-ray on the other, now compels the modeler to seek atmospheres models equal to the task, and not only those of solar composition. Thus, one of the most important new features for modern codes, in this editor's view - is the new set of atmospheres developed by Bob Kurucz, now incorporating 58 million lines. The binary stars group at the University of Calgary is now hard at work on simulating stellar flux to black-body ratios for the UBVRI, uvby, and various spectral bins in the intermediate and far ultraviolet, and is preparing to complete the job in the infrared, as these passbands become more standardized. The development of the modeling of atmospheric eclipses by stellar winds promises to be an important field in the future. Astronomers in the former Soviet Union have taken a leading role in this area and one of at least two Canadian-German groups studying massive stars is attempting to modify the *WD* program to model both type of eclipses. The trend in the wider community is toward further inclusion to make the models reflect more accurately the structural complexities presented by nature. Thus, the proceedings represent a collective statement of the condition of eclipsing binary modeling capability at the beginning of the last decade of the twentieth century, and point the way to future developments.

For all codes, improvements are clearly in the works. One of the more important aspects of light curve modeling is the availability of better atmospheric models, as Bob Kurucz has demonstrated very well. The question of uniqueness and determinacy, directly commented on by Dr. Cherepashchuk, has been examined too, and the tools now available to deal with these questions - such ILOT described by Banks and discussed by Budding, and the simplex enhancements described by Josef Kallrath - are shown to be formidable.

Acknowledgements

The aim was to have all final copies in hand by the end of 1991. With the cooperation of almost all of the presenters, this timetable was nearly met, resulting in a timely indication of the state of the art. It is regretted that the current volume does not contain the manuscripts of Prof. K.-C. Leung and Prof. E. F. Guinan, who, unable to complete the manuscripts by the deadline, graciously withdrew them rather than holding up publication of the proceedings.

In the interest of expediency on that final day in the Hotel Gran Dora, I had to omit my closing Spanish text, which refers to all participants, and expresses well-deserved thanks to all the Argentine organizers:

> *Muchas gracias a todos por su interés en este tema, y también a nuestros anfitriones por la oportunidad de reunirnos aqui y por haberse hecho cargo de organizar una reunión esta magnitud. Mis felicitaciones por la excelente tarea realizada.*

It is my agreeable task to thank Dr. R.H. Koch, President of Commission 42, and the other members of the present executive and Organizing Committee for their support for this meeting. Dr. Koch played several roles: advisor, intermediary, interpreter, recorder, and temporary chairman. In each role he was superb. The Scientific Organizing Commitee for these sessions consisted of the following: Dr. Graham Hill and myself of Canada; Dr. Zhang Di-Sheng of Beijing, China; Dr. Ron Hilditch of the U.K., and Dr. Robert E. Wilson and Dr. Kam-Ching Leung of the U.S. The meeting site in the Salón Turquesa of the Hotel Gran Dora was selected and paid for by the Local Organizing Committee of Colloquium No. 151. It is therefore a great pleasure to thank the chairmen of the organizing committees, Dr. R. F. Sisteró and Dr. Yoji Kondo for their support in enabling us to conclude the meeting satisfactorily following the unexpected events in Buenos Aires. It is a pleasure to thank all the participants at both segments of the meeting for helping to make the meeting so successful in spite of all exigencies. Gaston and Andrea Groisman provided most of the sensible Spanish syntax. Brief reports of the meeting proceedings were made for Comm. 42 by volunteer secretaries: Doug Hube of the University of Alberta and Bob Koch, who also chaired the meeting during my own presentation.

Finally, it is a pleasure to thank Dr. Tom von Foerster, Physics Editor at Springer-Verlag Publishers, New York, for his support for this project.

- E.F. Milone
Calgary, July, 1992

References

Carr, R. B., 1972. *AJ* **77**, 155.

Kallrath, J., Milone, E. F., and Stagg, C. R., 1992. *ApJ* **389**, 590.

Lepasso, E., 1992. In *Evolutionary Processes in Interacting Binary Stars*, R. Polidan, ed. (Dordrecht: Kluwer), *in press.*

Milone, E. F., Fry, D. J. I., and Grillmair, C., 1987. *BAAS* **19**, 643.

Milone, E. F., Groisman, G., Fry, D. J. I., and Bradstreet, D. H., 1991. *ApJ* **370**, 677.

Milone, E. F., Stagg, C. R., and Kurucz, R. L., 1992. *ApJS* **79**, 123.

Niarchos, P. G., 1978. *ApSpSci.* **58**, 301.

Rainger, R. P., Hilditch, R. W., and Bell, S. A., 1990. *MNRAS* **246** 42.

Samec, R. G., Van Hamme, W., and Bookmyer, B. B., 1989. *AJ* **98**, 2287.

Wilson, R. E., 1990, in *Report of IAU Commission 42*, ed., R. H. Koch, to appear in *Reports on Astronomy*, **XXIA** (Dordrecht: Kluwer), *in press.*